# Results and Prospects of Geobotanical Research in Siberia



Dedicated to the 75th Anniversary of the Laboratory of Ecology and Geobotany of CSBG SB RAS

BIO Web of Conferences Volume 16 (2019)

Novosibirsk, Russia
13 - 17 May 2019

**Editors:**

**N.N. Lashinsky**
**N.I. Makunina**

ISBN: 978-1-5108-9597-3

**Printed from e-media with permission by:**

Curran Associates, Inc.
57 Morehouse Lane
Red Hook, NY 12571

CURRAN ASSOCIATES INC. proceedings .com

**Some format issues inherent in the e-media version may also appear in this print version.**



For additional information, please contact EDP Sciences – Web of Conferences at the address below.

EDP Sciences – Web of Conferences
17, Avenue du Hoggar
Parc d'Activité de Courtabœuf
BP 112
F-91944 Les Ulis Cedex A
France

Phone: +33 (0) 1 69 18 75 75
Fax: +33 (0) 1 69 28 84 91

contact-edps@webofconferences.org

**Additional copies of this publication are available from:**

Curran Associates, Inc.
57 Morehouse Lane
Red Hook, NY 12571 USA
Phone: 845-758-0400
Fax: 845-758-2633
Email: curran@proceedings.com
Web: www.proceedings.com

# TABLE OF CONTENTS

edp sciences

BIO Web of Conferences

# Statement of Peer review

In submitting conference proceedings to BIO Web of Conferences, I certify to the Publisher that I adhere to the **Policy on Publishing Integrity** of the journal in order to safeguard good scientific practice in publishing.

1. All articles have been subjected to peer review administered by the proceedings editors.
2. Reviews have been conducted by expert referees, who have been requested to provide unbiased and constructive comments aimed, whenever possible, at improving the work.
3. Proceedings editors have taken all reasonable steps to ensure the quality of the materials they publish and their decision to accept or reject a paper for publication has been based only on the merits of the work and the relevance to the journal.

Title, date and place of the conference „Results and prospects of geobotanical research in "Siberian", dedicated to the 75th anniversary of the laboratory of ecology and geobotany of CSBG SB RAS (Russia, Novosibirsk, may 13-17, 2019)

Proceedings editor(s):

Lashinsky N.N. - chief researcher of laboratory of Ecology and geobotany, CSBG SB RAS, Dr.

Makunina N.I - senior researcher, laboratory of Ecology and geobotany, Dr.

Date and editor's signature

# A study of the Overburden dumps flora at Chernogorsky quarry

*Eugene* Ankipovich, *Elena* Lagunova*, *Irina* Barsukova, and *Tatiana* Leonova

Katanov Khakassia State University, Abakan, Russia, 655017

**Abstract.** Species composition of phytocoenoses of overburden dumps at Chernogorsky quarry has been investigated. Systematic, chorological, environmental and biomorphological analysis of flora has been carried out. It has been found that the quarry flora includes 47 species of higher vascular plants that are related to 35 genera and 16 families. 97.9 % are metasperms. *Asteraceae* and *Poaceae* are the largest families. Species with Holarctic areas prevail in the chorological structure (36.2 %). In terms of soil moisture all species can be subdivided into 3 groups: mesophytes (40.4 %), mesoxerophytes (36.2 %) and xerophytes (23.4 %). The leading role in the researched flora belongs to herbaceous polycarpous plants (57.4 %), which is characteristic of all the boreal region flora. 1 species listed in the Red Books of the Republic of Khakassia was discovered in the overburden dumps (*Nitraria sibirica* Pallas) and 6 species which have been noted for the first time for the flora of Khakassia.

## 1 Introduction

In case of open-cut mining enormous territories are subjected to harsh technogenic impact. Chernogorsky coal-field has been mined in Khakassia since 1956. By now, the total area of the ground damaged by open-cut mining is 3200 ha. Herewith, the territory that suffers from negative consequences related to the disturbance of hydrological regime, polluting adjacent landscapes with technogenic toxicants, increased erosion, etc. is 10 times bigger than the areas of the damaged land. Destruction of technogenic landscape by wind and dispersion of unconsolidated rock fractions at the overburden dumps are dangerous environment pollution sources as well. Transformation of natural landscapes of Khakassia into industrial and anthropogenic ones happens without sufficient landscape and environmental justification and clear forecast for decreasing negative consequences for the environment. In this connection, it is highly important to study the process of flora diversity recreation at overburden dumps of different age. The purpose of this study is to ascertain special features of natural colonization by vegetation of the overburden dumps at Chernogorsky quarry. In accordance with this purpose, we set the objective to discover and analyze the species composition of phytocenoses, confined to microrelief plots of uneven-aged overburden dumps.

* Corresponding author: aes-1962@yandex.ru

## 2 Materials and methods

The research materials are the samples of plants collected during field trips in 2018-2019. The main research method was transecting (Transect method), covering most coal overburden dumps at Chernogorsky quarry. The gathered collections were handed over to the Herbarium of Khakass State University (HGU). The authors would like to express their gratitude to D.N. Shaulo, the Head of the Herbarium Lab of the Central Siberian Botanical Garden, SB RAS, for the kind help and consultations in the process of plant determination.

## 3 Results and their discussion

The enterprise SUEK Khakassia LLC, Chernogorsky quarry was formed on the basis of Chernogorsky hard coal field, located in the territory of Ust-Abakansky district of the Republic of Khakassia. The monitoring results demonstrate that the area of disturbed soils has increased from 2061 ha to 3168 ha over the last 25 years [1]. The studied overburden dumps are of external type and comprise a ravine stretching from East to West, which widens in the western part and is overlapped with transverse dump of ridge-like surface. The ravine height above the sea level is 402 m, the height of the top of the overburden dump is 425 m. 5 elements of the mesorelief stand out: plateau, the slopes of the Northern, Souhern, Western and Estern expositions. In the Western part of the lot, the ravine is complicated by smaller ravines, small closed topographic lows and hills [2]. In terms of geological structure, it is comprised of mid-carboniferous coal-bearing strata of Chernogorsky formation and modern technogenic strata. Soil-forming rock is virtually absent on the slopes and the surface of the dumps [3]. The quarry is located in Minusinsk Hollow, which is characterised by anticyclonic conditions of atmospheric circulation with the prevalence of dry, mostly cloudless weather, sudden temperature fluctuations and interchanging periods of light breezes and strong winds [4].

The flora of overburden dumps at Chernogorsky quarry includes 47 species of higher vascular plants, related to 35 genera and 16 families. By 97.9 % the flora consists of angiosperms; 74.5 % of these are dicotyledonous, 23.4 % are monocotyledonous. Vascular sporophytes are represented by one specie *Pteridium aquilinum* (L.) Kuhn. The largest families within the researched flora are *Asteraceae,* which includes 12 species (25.5 % of the flora composition) and *Poaceae,* which includes 11 species (23.4 %). The most dispersed species of the *Asteraceae family*: *Artemisia frigida* Willd., *A. glauca* Pall. ex Willd., *A. sieversiana* Willd., *Heteropappus altaicus* (Willd.) Novopokr., *Erigeron canadensis* L. The high plastisity of the species of this family, and the capacity to adapt to unfavourable conditions gives them the possibility to occupy new habitats easily, including the anthropogenic ones. The *Poaceae* reinforce the soils of overburden dumps at the abovementioned quarry. The flowing species are widely represented: *Calamagrostis epigeios* L., *Leymus racemosus* (Trin.) K. Richt., *Setaria viridis* (L.) P. Beauv., *Elytrigia repens* L., *Hordeum jubatum* L.

*Fabaceae*, *Rosaceae* и *Brassicaceae* families contain 3 species each (6.4 %). Species of *Fabaceae* family are usually abundant in the steppe floras of Siberia, however, legums are scarce on overburden dumps. Synanthropic species *Onobrychis arenaria* (Kit.) DC. and *Melilotus officinalis* (L.) Pall have been discovered here. These species' biomass increases and they are able to invade large territories quickly in conditions of ruderal ecotopes, especially if they are part of a pioneer group. *Astragalus atrosibiricus* Schischkin is rarely found in the researched territory. *Rosaceae* family in the disturbed flora of overburden dumps at Chernogorsky quarry is determined by its adventive species: *Physocarpus opulifolius* (L.) Maxim. and *Potentilla lydiae* Kurbatsky. The specie *Potentilla tanacetifolia* Willd. ex Schlecht is encountered rather frequently. In the same way as *Fabaceae*,

*Brassicaceae* family is represented by synanthropic species, which are spread with seeding material in ruderal habitats and includes the following species: *Erisimum cheiranthoides* L., *Sinapis arvensis* L., *Sisymbrium wolgense* M. Bieb. ex E. Fourn.

*Chenopodiaceae*, *Scrophulariaceae*, *Solanaceae* and *Caryophyllaceae* families are represented by 2 species each (4,3 %). 6 families (*Primulaceae*, *Lamiaceae*, *Boraginaceae*, *Nitrariaceae*, *Onagraceae*, *Gypolepidaceae)* contain 1 specie each in the researched flora.

The analysis of the generic spectrum demonstrated that there is one multi-specific genus on the slopes of uneven-aged overburden dumps, *Artemisia*, which includes 8 species (17 %) (*Artemisia dracunculus* L., *A. scoparia* Waldst. et Kit., *A. frigida*, *A. macrocephala* Jacq. ex Bess., *A. sieversiana*, *A. glauca*, *A. tanacetifolia* L., *A. vulgaris* L.). The genera *Potentilla, Leymus, Poa, Linaria, Solanum* contain 2 species each (4.3 %). The other genera include 1 specie each. The large proportion of homotypical and species-limited genera point to the relative newness of the flora. The homotypical and byspectral genera cover 82,9% of the of the generic diversity. 61.7 % of all the genera under study are homotypical. According to Tolmachev, A.I. [8], the prevalence of species-limited and, especially, homotypical genera in the particular flora can point to the fact that the given flora was to a great degree formed due to migration of species from the adjacent territories.

The chorological structure consists of 6 groups. Holarctic species prevail (17 species, or 36.2 % from the overall number of species). The Eurasian group is represented by 13 species (27,6%), including 3 cosmopolitan species (6.4 %). The share of the species which do not spread beyond Asia (North-Asian, Central Asian, South Siberian) is 29.8 % of all the species, which is the evidence of allochthonous trends in the flora development.

Groups of plants with different relation to soil moisture are marked out for environmental analysis [5]. All the species are related to 3 ecological groups. The greatest number of species is found in the mesophytes group (19 species or 40.4 %), including *Cirsium setosum* (Willd.) Bieb., *Inula salicina* L., *Convolvulus arvensis* L., *Androsace septentrionalis* L., *Linaria vulgaris* Mill., *Melandrium album* (Mill.) Garcke, *Sinapis arvensis* L., *Chamerion angustifolum* (L.) Holub., *Pteridium aquilinum* (L.) Kuhn, *Physocarpus opulifolius* (L.) Maxim., etc. Mesoxerophytes include 17 species (36.2 %). These are such species as *Potentilla tanacetifolia* Willd. ex Schlecht., *Onobrychis arenaria* (Kit.) DC., *Heteropappus altaicus*, *Elytrigia repens, etc.* Xerophytes are represented by 11 species (23.4 %). *Linaria altaica* Fisch., *Nitraria sibirica* Pall., *Koeleria cristata* (L.) Pers., *Artemisia frigida,* etc. belong to this ecological group.

The analysis of life forms has been carried out on the basis of I. G. Serebryakov classification [6]. In the flora under study, just as in the flora of Khakassia [7] and in the flora of the boreal region in general [8], the leading role belongs to herbaceous polycarpous plants. They comprise 57.4 % of the overall number of species. The main part of perennial grasses belongs to the group of taproot plants and includes 8 species (17 %): *Onobrychis arenaria*, *Astragalus austrosibiricus* Schischkin, *Lithospermum officinale* L., *Linaria vulgaris* Mill., etc. There are 7 species of long-rooted plants (23.6 %), including mostly mesophyllous species *Chamerion angustifolum*, *Inula salicina, Artemisia tanacetifolia, Cirsium setosum, Convolvulus arvensis* L. Short-rooted plants are represented by 6 species (12.8 %) (*Potentilla tanacetifolia*, *P. lydiae*, *Artemisia glauca*, *A. dracunculus* L., *Pteridium aquilinum* (L.) Kuhn, *Sisymbrium wolgense*). Sod plants annually open not one but a great number of reproduction buds therefore large "bushes" or groves are formed [6]. Sod plants are subdivided into densely bushy and bunch-grasses, with 3 species of each (6,4%) and belong to *Poaceae* family. *Koeleria cristata*, *Psathyrostachys juncea* (Fisch.) Nevski, *Poa botryoides* (Trin. ex Griseb.) Roshev are densely-bushy plants. *Leymus dasystachys* (Trin.) Pilger**,** *Poa angustifolia* L., *Elytrigia repens* are bunch-grasses.

Monocarpous herbs include 16 species, which comprises 34 % of the overall number of species. Many of them are weeds. The most multiple representatives of annual and biennial plants belong to the following families: *Poaceae* (*Hordeum jubatum*, *Panicium miliaceum*, *Setaria viridis*), *Lamiaceae* (*Leonurus tataricus* L.), *Caryophyllaceae* (*Melandrium album* (Mill.) Garcke), *Brassicaceae* (*Erisimum cheiranthoides* L., *Sinapis arvensis* L.), *Primulaceae* (*Androsace septentrionalis*), *Chenopodiaceae* (*Chenopodium aristatum*, *Salsola collina* Pallas.), *Solanaceae* (*Solanum nigrum* L.), *Asteraceae* (*Erigeron canadensis*, *Artemisia sieversiana*, *A. scoparia* Waldst. et Kit.), *Fabaceae* (*Melilotus officinalis*). Arboreal and semiarborescent plants are scarce in the researched flora: 2 shrubs (4,3%), *Physocarpus opulifolius* and *Nitraria sibirica*, subshrub *Solanum dulcamara L.* and dwarf subshrub *Artemisia frigida*.

One species from the Red Books of the Republic of Khakassia (*Nitraria sibirica*) has been discovered in the flora of overburden dumps at Chernogorsky quarry [9] and six species, which have not yet been observed in Khakassia before, including *Physocarpus opulifolius, Potentilla lydiae, Artemisia macrocephala* Jacq. ex Bess., *Leymus dasystachys* (Trin.) Pilger, *Sisymbrium wolgense, Solanum dulcamara* have also been discovered.

The work was financially supported by the grants of the Russian Foundation for Basic Research №18-44-190006 and №18-44-190007.

## References

1. I.V. Zenkov, B.N.Nefedov, Yu.P. Yuronen, I.M. Baradulin, E.V. Kiryushina, V.N. Vokin, Coal., 2 (2015)
2. O.S. Safronova, T.G. Lamanova, N.V. Sheremet, Coal., 7 (2018)
3. I.V. Zenkov, B.N. Nefedov, I.M. Baradulin, V.N. Vokin, E.V. Kiryushina, Ecology and the industry of Russia, 6 (2014)
4. A.V. Kuminova, N.M. Chizhikova, M.G. Tanzybaev, *The vegetation cover of Khakassia* (Nauka, Novosibirsk, 1976)
5. T.G. Goryshina, *Ecology of plants* (Vysshaya Shkola Publishers, Moscow, 1979)
6. I.G. Serebryakov, *Ecological morphology of plants. Life forms of angiosperms and coniferous plants* (Vysshaya Shkola Publishers, Moscow, 1962)
7. E.S. Ankipovich, *The catalogue of the flora of the Republic of Khakassia* (Altay University Publishing House, Barnaul, 1999)
8. A.I. Tolmachev, *Introduction into plants geography* (LSU Publishing House, Leningrad, 1974)
9. *The Red Books of Khakassia. Rare and endangered species of plants and mushrooms* (Nauka Publishing House, Novosibirsk, 2012)

BIO Web of Conferences **16**, 00002 (2019) https://doi.org/10.1051/bioconf/20191600002
*Results and Prospects of Geobotanical Research in Siberia*

# Taxonomic structure of local floras in Sidinskiy and Pribaitakskiy steppes (Krasnoyarsk region)

*Ekaterina* Antipova*, *Olga* Enulenko

Krasnoyarsk State Pedagogical University named after V. P. Astafiev, Ada Lebedeva Str., 89, Krasnoyarsk, 660049, Russia

**Abstract.** Sidinskiy and Pribaitakskiy steppes are the part of the Minusinsk province of the Altai-Sayan mountain country. On the territory of steppes 12 local floras (LF) were marked and examined using the same method. Studies were carried out according to the recommendations of A.I. Tolmachev (1931): identification of different types of ecotypes of LF, the full species composition of ecotypes (flora inventory), tracking the completeness of detection in LF. In the Sidinskiy steppe the number of families in the LF varies from 25 (Syida) to 43 (v. Alha), in the Pribaitakskiy steppe there is a big difference in the number of families in the LF from 31 (Krasnoyarsk reservoir, Tuba) to 65 (Mayak, Idra). The leading families in the local steppe flora are Asteraceae, Poaceae, Fabaceae. The Fabaceae family ranks third in the top of ten families in 50 % of LF, determining the type of flora as the Mediterranean-Central Asian (Fabaceae-type). Different types of LF allow to determine the type of all steppe flora as mixed, combining features of boreal (Asteraceae-Poaceae-type) and steppes flora (Fabaceae-type) with close ties with Central European floras (Rosaceae-type). 4 distinct levels of species richness are determined: 1 – 496-544; 2 – 408-413; 3 – 319-374; 4 – 279-298 species.

Sidinskiy and Pribaitakskiy steppes are located in the Sydo-Yerbin Basin, are a part of the Minusinsk province of the Altai-Sayan mountain country [1]. Sidinskiy steppe occupies the southern part of Krasnoturanskiy region, Pribaitakskiy steppe covers a part of Krasnoturanskiy, Idrinskiy and Kuraginskiy areas of the Krasnoyarsk region.

On the territory of steppes 12 local floras (LF) were marked and examined using a single method. These LF are samples of flora [2-4] and are representing the main variety of partial floras in the vicinity of the studied geographical points. Studies were carried out according to the recommendations of A. I. Tolmachev [5], which implies equal completeness of the study of local floras: identification of different types of ecotypes of LF, the full species composition of ecotypes (flora inventory), tracking the completeness of detection in LF.

Comparing studied steppes LF, revealed different levels of floristic richness associated with landscape, environmental and edaphic features, as well as the difference in their areas. Pribaitakskiy steppe is superior then Sidinskiy one by area and the number of studied local floras: 3 LF – in Sidinskiy and 9 in Pribaitakskiy steppes (table 1).

* Corresponding author: katusha05@bk.ru

**Table 1.** Parameters of local floras in Sidinskiy and Pribaitakskiy steppes

| LF | The composition of the flora | | | 10 leading families of LF | | Monospecific families of LF | |
|---|---|---|---|---|---|---|---|
| | familie s of LF | genus of LF | species of LF | the absolute number of species | percent compositio n of the total number of species | the absolut e number of species | percent compositi on of the total number of species |
| Sidinskiy | | | | | | | |
| **Al** | 43 | 202 | 349 | 228 | 65,3 | 17 | 1,7 |
| **Un** | 42 | 122 | 413 | 245 | 59,3 | 14 | 1,4 |
| **Sd** | 25 | 59 | 279 | 159 | 57,0 | 10 | 1 |
| **Pribaitakskiy steppe** | | | | | | | |
| **Mk** | 65 | 272 | 544 | 378 | 69,4 | 21 | 2,1 |
| **Kb** | 49 | 139 | 419 | 241 | 57,5 | 20 | 2 |
| **Tu** | 64 | 237 | 496 | 406 | 81,8 | 27 | 2,7 |
| **Lo** | 33 | 74 | 298 | 221 | 74,2 | 17 | 1,7 |
| **Id** | 65 | 232 | 408 | 366 | 89,7 | 29 | 2,9 |
| **Bs** | 37 | 102 | 368 | 218 | 59,2 | 11 | 1,1 |
| **Ku** | 47 | 183 | 365 | 222 | 60,8 | 19 | 1,9 |
| **Kv** | 31 | 111 | 374 | 225 | 60,1 | 9 | 0,9 |
| **Tb** | 31 | 95 | 316 | 243 | 76,9 | 3 | 0,3 |
| **In total:** | 102 | 365 | 994 | 603 | 60,3 | 38 | 3,8 |

**Note**. LF – local floras; Al – v. Alha, Un - v. Unyuk, Sd - Syida, Mk – city Mayak, Kb – Krasnoturanskiy Bor, Tu – city Turan, Lo – lake Lebyazhye, Id – Idrinskoe, Bs – Big Salba, Ku – Kuragino, Kv – Krasnoyarsk water reservoir, Tb – Tuba.

The largest number of families and genus are represented in local floras with a combination of flat and mountain landscapes. In the Sidinskiy steppe the number of families in the LF varies from 25 (Syida) to 43 (v. Alha), the richest LF of Unyuk. An equally large difference in the composition of the steppe genus (59-202) is explained by local orography conditions – the presence of v. Alha, Unyuk, river Syida with its valleys and the Syidin Bay of the Krasnoyarsk water reservoir. Syida local flora is isolated by the Syidin Bay and includes a large number of intrazonal vegetation communities, therefore, contains the smallest number of families and genus (table 1).

In the Pribaitakskiy steppe there is a big difference in the number of families in the LF from 31 (Krasnoyarsk reservoir, Tuba) to 65 (Mayak, Idra) and genus (from 74 (lake Lebyazhye) to 272 (Mayak), due not only to the hilly-flat terrain, rich network of rivers with swampy depressions, where steppe, meadow, water and wetland vegetation is the most fully represented, but also a combination of pronounced latitudinal zoning and altitudinal zoning, manifested in concentric zoning. In the Central part of the Pribaitakskiy steppe local flora of Turan, Mayak, Krasnoturanskiy bor forms the "steppe core", surrounded by meadow steppes wide striped on the periphery.

The leading families in the local steppe flora are Asteraceae, Poaceae, Fabaceae. In taxonomic spectras 12 LF (table 2) the coincidence of the set of leading families is a characteristic only for 2 LF. All LF take first and second place of the family Asteraceae and Poaceae, except for the LF Syida, where they change their places. These are the largest families for the flora of the entire globe, as well as for all areas of the Holarctic (with the exception of the Arctic regions) [6]. This pattern is reflected in all local floras studied steppes.

The Fabaceae family ranks third in the top of ten families in 50 % of LF (Alha, Turan, lake Lebyazhye, Kuragino and Krasnoyarsk water reservoir) with high species richness (17-39), determining the type of flora as the Mediterranean-Central Asian (Fabaceae-type). This is a manifestation of the ancient "southern" links with the formation of the secondary center of species spreading in the steppes [6-9]. The remaining LF belong to different types: 3 LF (Maiak, Krasnoturansky Bor, Idra) are characterized by close ties with the Central European flora (Rosaceae-type) with the manifestation of features of climate humidity. The local flora of Tuba bears the features of the middle Eastern and Eurasian floras with a hilly meadow, forest-meadow and meadow-marsh gradient (Ranunculaceae-type). Local flora Big Salba is defined as Arctoboreal-East-Asia-combined, which combines features of boreal and steppe floras. The local flora of Unyuk bears the features of Alpine-meadow flora with the European bias and influence of species of anthropogenic flora (Brassicaceae-type).

Thus, the families of the leading part of the taxonomic spectrum of LF are associated with their zonal and subzonal position, as well as the alternation of humid and arid habitats in the steppes. The set of families leading by the number of species corresponds to the geographical position of the steppes, reflecting the main features of the boreal flora (predominance of families Asteraceae, Poaceae, Rosaceae, Ranunculaceae, Cyperaceae) with an ancient Mediterranean bias. High ranks make up the family (Fabaceae, Lamiaceae, Apiaceae) due to migration of species from neighboring floristic regions of Kazakhstan and Central Asia. In the majority of the LF mixed migratory species are dominating. The relative abundance of Eurasian species increases to the Northeast and East of the studied steppes.

**Table 2.** Leading families of floristic list of local floras Sidinskiy and Pribaitakskiy steppes

| Families | Local floras | | | | | | | | | | | |
|---|---|---|---|---|---|---|---|---|---|---|---|---|
| | Al | Un | Sd | Mk | Kb | Tu | Lo | Id | Bs | Ku | Kv | Tb |
| *Asteraceae* | 54/1 | 42/1 | 34/2 | 138/1 | 97/1 | 72/1 | 36/1 | 86/1 | 57/1 | 87/1 | 37/1 | 46/1 |
| *Poaceae* | 42/2 | 39/2 | 49/1 | 63/2 | 73/2 | 43/2 | 28/2 | 44/2 | 43/2 | 58/2 | 25/2 | 38/2 |
| *Fabaceae* | 20/3 | 20/4 | 16/3 | 38/4 | 22/4 | 39/3 | 17/3 | 29/4 | 15/5 | 29/3 | 17/3 | 15/5 |
| *Rosaceae* | 17/6 | 11/6 | 10/6 | 43/3 | 34/3 | 34/4 | 11/4 | 32/3 | 14/7 | 15/5 | 10/8 | 9/7 |
| *Ranunculaceae* | 18/5 | 18/5 | 14/5 | 33/5 | 19/5 | 19/6 | 9/7 | 22/6 | 17/4 | 14/6 | 9/9 | 36/3 |
| *Brassicaceae* | 19/4 | 22/3 | 9/7 | 33/6 | 12/8 | 19/7 | 9/8 | 22/7 | 15/6 | 14/7 | 15/4 | 9/8 |
| *Lamiaceae* | 12/7 | 10/8 | 5/8 | 16/9 | 14/7 | 23/5 | 4/9 | 18/10 | 9/9 | 14/8 | 15/5 | 8/9 |
| *Polygonaceae* | 8/8 | 11/7 | 5/9 | 16/10 | 12/9 | 15/8 | -- | 20/9 | -- | -- | 12/6 | -- |
| *Cyperaceae* | 6/10 | 8/10 | 16/4 | 26/7 | 16/6 | 8/10 | 8/6 | 23/5 | 18/3 | 14/9 | 12/7 | 27/4 |
| *Apiaceae* | 7/9 | 9/9 | -- | 18/8 | 5/10 | 12/9 | 11/5 | 21/8 | 12/8 | 17/4 | 9/10 | 10/6 |

**Note**. Absolute number of vascular plant species in the family/ number of family place descending.

Different types of LF allow to determine the type of all steppe flora as mixed, combining features of boreal (Asteraceae-Poaceae-type) and steppes flora (Fabaceae-type) with close ties with Central European floras (Rosaceae-type).

The amplitude and the course of change in the number of LF species in the compared steppes is different. 4 distinct levels of species richness are determined: 1 – 496-544; 2 – 408-413; 3 – 319-374; 4 – 279-298 species (table 2).

**Fig.** Species diversity in local floras Sidinskiy and Pribaitakskiy steppes (in the circles is the number of species). Decoding of local floras names is given in table 1.

In the Sidinskiy steppes on the level of species richness allocated LF Alha and Unyuk, in the Pribaitakskiy steppes -LF Mayak, Idra, Krasnoturanskiy Bor, according to the nature of the vegetation cover attributed by L. M. Cherepnin [10] to the subzone meadow steppes (southern steppes), which combines steppes, forest-steppes and meadow species. The value of more than 500 species (544) was observed only in the Central border point between Sidinskiy and Pribaitakskiy steppes – in the local flora Mayak.

The level of species diversity increases from North to South and from West to East and North-East, as well as from the periphery to the center of the studied flora (Illustration) that is associated with the increase in the continentally of climate in these areas, concentric zoning, which often overlaps with macromosaic, a variety of antopia and enclaves of "southern" vegetation.

Acknowledgement: the Study was carried out with the financial support of the Russian Foundation for basic research, the Government of the Krasnoyarsk territory, the Krasnoyarsk regional Science Foundation in the framework of the scientific project 18-44-240006 "Natural and urbanized flora of the Yenisei Siberia".

## References

1. Yu.A. Sherbakov, M.V. Kirillov, Sib. geographer. team, **1** (1962)
2. B.A. Yurtsev, Bulletin of MOIP, **87**, 4 (1982)
3. E.M. Antipova, *The flora of the Northern forest-steppes of Middle Siberia*, Diss. (Tomsk, 2008)
4. E.M. Antipova, Comparative Floristic: Materials of all-Russian school-seminar on comparative Floristic, **2**, 2 (2010)
5. A.I. Tolmachev, Journal. Russ. bot. Ls-VA, **16**, 1 ( 1931)
6. A.P. Hohryakov, Bot. journal, **85**, 5 (2000)

7. A.V. Polozhii, *Flora of Krasnoyarsk region: Fabaceae*, **6** (Tomsk, Tomsk state University; Novosibirsk, Science, 1960)
8. A.I. Tolmachev, *Introduction to the geography of plants* (L., publishing House of LSU, 1974)
9. A.I. Tolmachev, *Methods of comparative Floristics and problems of florogenesis* (Novosibirsk, Science, 1986)
10. L.M. Cherepnin, *Flora and vegetation of the southern part of Krasnoyarsk region*, Diss. (L., BIN AN SSSR, 1953)

# Coenopopulation structure of some high-mountain *Nepeta* species (serias *Densiflora)* in Central Asia

*Alexey Yu.* Astashenkov*

Central Siberian Botanical Garden of the SB RAS, Zolotodolinskaya Str., 101, Novosibirsk, 630090, Russia

**Abstract**. The analysis of coenopopulations of high-mountain *Nepeta* ser. *Densiflorae* from Central Asia is conducted. It is established that changes in the peaks of the ontogenetic spectrum from centered to double-peak centered reflect change of the biomorph from tap-root caudex to long-rhizome-tap-root (caudex). The biomorph transformation directs at a gradient of changing environmental conditions: from cryoxerophitic to mesophitic habitats.

## 1 Introduction

In the process of evolution, the plants of high-mountainous belts produced various mechanisms of adaptations, the combination of which provided them with a steady modern existence. Group of such plants includes most species of the genus *Nepeta* from the Lamiaceae family. Perennial *Nepeta* have mainly endemic areas associated with alpine orogenesis, which influenced on their distribution and evolutionary development in the highlands of Central Asia. A comprehensive analysis carried out at various levels of the organization makes it possible to identify phylogenetic relationships and the main directions of evolutionary transformations at the population, species, and supraspecific (within closely related groups) levels.

The population-ontogenetic approach [1, 2], based on an assessment of the organismic and population levels and widely used in modern biology, makes it possible to determine the state of natural coenopopulations (CP) under various ecological-phytocoenotic conditions [3]. Population analysis of the *Nepeta* species and search for relationships between the elements of the morph and phytocoenotic environment were not conducted.

The purpose of the work is to identify the features of the population structure of some species of *Nepeta*. High-altitude *Nepeta* of different biomorphs, the taxa of which are united by a related group of the polymorphic series *Densiflorae* Pojark., were chosen as objects of study.

* Corresponding author: astal@bk.ru

## 2 Material and methods

We used classical population-ontogenetic approaches. When studying ontogenesis, the concept of its discrete description was adopted [4–6]. The morphological study of the individuals was conducted both in the field and in the laboratory – on a fixed (herbarium) material (mainly). Ontogenetic spectra of coenopopulations were built on the basis of recording individuals on 25 sites of 1 $m^2$ in size, laid in a regular way on transects 1m wide in each coenosis. An individual (genet) and partial formations (ramet) are taken for the counting unit. The characteristic of ontogenetic structure of coenopopulations is given according to generally accepted methods [1, 2]. The classification of "delta-omega" by L.A. Zhivotovsky [7] was used for determine the type of coenopopulation. The following demographic indicators: $\Delta$ – age index [5] and $\omega$ – efficiency index were taken [7]. Ecological density was established on the basis of the number of individuals per inhabited space unit (ind./$m^2$) [8].

## 3 Results and discussion

*N. kokamirica* Regel – Tash-Shan-Dzungar-Kashgar herbaceous, caudex-forming perennial. The species grows on stony and gravelly mountain slopes, often on mobile scree substrates. It enters into the composition of the vegetation from the upper forest to the alpine zone [9]. The material was collected on the southern macroslope of the Dzungarian Alatau: Koksu Range, the gorge of the Burkhan river, semi-fixed loess-gravel scree, the upper border of the sub-alpine belt – a primitive plant aggregation of prostrate juniper stand (*Juniperus pseudosabina* Fisch. et C. Mey.). The spatial distribution of individuals at survey sites in the absence of phytocoenotic pressure has uneven group arrangement – min 1–2 ind./$m^2$, max 15–17 ind./$m^2$. Individuals by local groups are concentrated along the upper edge of the axial part of the mountain slope in an open area surrounded by *J. pseudosabina*. The density of individuals in the CP was 5.2 ind./m2. Ontogenetic spectrum is centered, with an absolute maximum for mature generative plants. That type of spectrum is associated with the duration of this ontogenetic state. According to the classification of L. A. Zhivotovsky [7], the coenopopulation is aging: the age index ($\Delta$) is 0.56, the efficiency ($\omega$) –0.65. The obtained values reflect the high number of old generative plants and individuals of the post-generation period. Nevertheless, the state of the studied CP, with high values of mature generative plants, can be characterized as normal, stable.

*N. pulchella* Pojark. – short-rhizome-tap-root (caudex) herbaceous perennial. Endemic of the Talas Alatau (Western Tien Shan). It grows in the composition of meadow and herb-grass-meadow coenoses on the northern macroslope of the ridge from the middle mountains to the subalpine belt [9]. The CP was studied in the Ulken-Kaindy gorge in the highlands. A single-peaked centered ontogenetic spectrum is formed in this CP. That type is associated with the continuous development of the individuals in the generative period, which coincides with the time of formation of the short rhizome. Particulation of the bush and the formation of long-lived ramet also influence on the delay in the rate of development of individuals. A significant proportion of old generative individuals are due to the re-partitioning of the daughter ramet, which also leads to extended ontogenesis. CP is transition to mature: $\Delta$ – 0.50, $\omega$ – 0.65. Ecological density is 3.9 ind./$m^2$.

*N. densiflora* Kar. et Kir. – long- rhizome-tap-root (caudex) herbaceous perennial, distributed in the southern part of Gorny Altai. The border of the distribution is limited to Mongolia and East Kazakhstan. The species is confined to the mesophilic highlands of the alpine mountain belt. It includes in the composition of rarefied plant aggregation of screes and rocky mountain slopes [9]. The material was collected on the plain part of the Narym Range in the headwaters of the river Jajdak in the erozion sleeve of a temporary watercourse embedded in *Alchemilla* meadow The spatial distribution of individuals in the absence of phytocenotic influence has an uneven group arrangement from 2 to 19 ind./$m^2$.The individuals concentrated along the edges of the temporary water flow sleeve. The density of individuals averaged 3.8 ind./m2. Ontogenetic spectrum is two-vertex centered. The absolute maximum in the spectrum turned out to be on juvenile individuals, the second – on the mature ones. A significant proportion of young individuals is associated with vegetative and seed regeneration. Nevertheless, the distribution of ontogenetic groups in CP reflects the characteristic ontogenetic spectrum of perennial plants of the root-root biomorph [2]. The CP according to the classification of L.A. Zhivotovsky is young: $\Delta$ – 0.24, $\omega$ – 0.46.

## 4 Conclusion

An analysis of the ontogenetic structure of the coenopopulations of studied alpine species showed a vector of change in the ontogenetic structure associated with a change in the biomorph. The rearrangement of the ontogenetic spectrum peaks from centered to double-top centered reflects the change of biomorph from tap-root caudex to long- rhizome-tap-root (caudex) and directs in a gradient of changing environmental conditions from cryoxerophilic to mesophilic habitats. Various mechanisms of self-maintenance of the studied species CP determine their sustainable development in the plant communities of Central Asia.

The work was carried out according to the draft State assignment of the Central Siberian Botanical Garden of the Siberian Branch of the Russian Academy of Sciences No. AAAA-A17-117012610053-9.

## References

1. *Plant coenopopulations: general terms and structure* (M, Nauka, 1976)
2. *Plant coenopopulations: basic concepts and structure* (M, Nauka, 1988)
3. L.B. Zaugol'nova, *Structure population of seed plants and related monitoring problems*, Diss. (St. Petersburg, 1994)
4. T.A. Rabotnov, Tr. Bot. Inst., Akad. Nauk SSSR, **6**, 3 (1950)
5. A.A. Uranov, Biol. Sciences, 2 (M,1975)
6. L.A. Zhukova, *Population Life of the Meadow Plants* (Yoshkar-Ola, Lanar, 1995)
7. L.A. Zhivotovsky, Rus. J. Ecol., 1 (2001)
8. Yu. Odum, *Ecology*, **2** (M, Nauka, 1986)
9. *Flora of the Soviet Union*, **20** (M.-L., Nauka, 1954)

# Biological peculiarities and characteristics of *Erodium tataricum* WILLD. cenopopulation in Khakasia

*Irina* Barsukova*, and *Tatiana* Leonova

Khakas State University n.a. N.F. Katanov, 655017, Russia

**Abstract.** Biological and structural peculiarities of *Erodium tataricum* Willd. cenopopulations in natural surroundings of Khakasia are examined. It is found out that the species make two vegetal forms under the changing environmental conditions. Ontogeny is complete, regeneration does not take place. Two types of ontogeny are found out: morphological and dynamic (according to the rate of growth). Examined *Erodium tataricum* Willd. cenopopulations are normal and incomplete. They are characterized by left-sided and bimodal ontogenetic spectra.

## 1 Introduction

*Erodium tataricum* Willd. (family *Geraniaceae* Juss.) – perennial herbaceous plant has a status of 3(R) in the Red List of Threatened Species of Russian Federation (2008) and in the Red List of Threatened Species of Khakasia (2012). It is paleoendemic, the area of species does not exceed the boarders of Khakasia [2]. Its basic area is between the rivers Belyi Iyus and Son in forest-steppe and steppe zone on stony and rank slopes. In connection with this, the biological and structural examination of the rare species of its cenotic population will allow to predict further development of cenopopulation and to take special activities for its protection. There is very little information about *E. tataricum* biology. The goal is to examine biological and structural peculiarities of *E. tataricum* cenopopulations in natural surroundings of Khakasia.

## 2 Material and methods

Cenopopulations (CP) were described in Minusinsk basin on the territory of Iyuso-Shirinskiy steppe in 2016-2018. Geobotanic descriptions were performed according to the established methodology [3]. The concept of its discreet description was taken into account when studying the ontogeny [4-7]. Biometric characteristics of species was made on the basis of the following features: the height of plants, the number of leaves and their length, the number of vegetal and generative shoots, the number of flowers, the length and diameter of caudex. The total number of *E. tataricum* specimens under study was 89.

* Corresponding author: saphronovairina@mail.ru

Ontogenetic structure of CP was performed with the use of population genetic approach [8-9]. 16-20 fixed areolas 1 $m^2$ in size were put in transects, all species were mapped with marking their ontogenetic condition. A species was taken as a unit of time. Understanding of L.B. Zaugolnova (1994) served as a basis when building the ontogenetic spectra and structure characteristics of CP. Demographic indices were used: age index (Δ) and efficiency index (ω) [10]. Space structure was examined according to the methodology of Y. Odum (1986).

**CP 1** was described in Shirinskiy region 12-15 m. away from the village Black Lake along the right bank of the river Belyi Iyus on the South-West slope of the hill in Karaganovaya motley grass-grasses petrophytic true steppe. The total projective cover (TPC) of herb layer was 40-45% (*Poa stepposa* (Krylov) Roshev., *P. botryoides* (Trin. ex Griseb.) Kom., *Festuca valesiaca* Schleich. ex Gaudin, *Artemisia frigida* Willd., *Potentilla acaulis* L.). Shrubbage accounts for 15-20%. The dominant position occupied *Caragana pygmaea* (L.) DC. *Dasiphora fruticosa* (L.) Rydb. and *Cotoneaster melanocarpus* Fisch. ex Blytt were growing sporadically in hollows.

**CP 2** was examined on the border with the State nature reserve «Khakassky» («The Lake Itkul», Shirinsky region) on the South-East slope of the hill in Karaganovaya forb-fecue-feather grass lithophyte true steppe (*Caragana pygmaea*, *Festuca valesiaca*, *Koeleria cristata* (L.) Pers., *Orostachys spinosa* (L.) C. A. Mey., *Potentilla acaulis*). The total projective cover (TPC) of herb layer was 50-55%. Projective cover of species in plant community made up 1%.

## 3 Results and discussion

According to the performed examination, *E. tataricum* is grown in similar plant communities. The dominant species are *Festuca valesiaca*, *Koeleria cristata*, *Poa stepposa*, *P. botryoides*, *Carex duriuscula* C. A. Mey., *Carex pediformis* C. A. Mey. *Orostachys spinosa*, *Potentilla acaulis*, *Artemisia frigida*, *Alyssum lenense* Adams, *A. obovatum* (C.A. Mey.) Turcz., *Thymus serpyllum* L., *Galium verum* L., *Pulsatilla reverdattoi* Polozhij & Malzeva and others accompany it.

*E. tataricum* – herbaceous taproot multi-headed polycarpous plant with assimilated rosette-like shoots of non-succulentus type. 3-5 (rarely 6-8) rosette-like shoots are formed on plant of a mature genesic ontogenetic condition. Their maximum height is 10 cm (more often 4-6 cm). The leaves are macropodous and pinnatifid. Their number on one vegetative shoot is 5-7. 2-4 genesic shoots 9 cm in height are developed in sinuses, each bearing 2-4 flowers. The plant is wolly, with white long pressed hair with a touch of grandular hair (specifically on flower-bearing stem and a flower cup). The presence of caudex make a species static – monocentric. Caudex multi-headed forms 3-9 branches 4,0-4.8 cm in height. The diameter of caudex in its expanded part varies from 0,9 to 1.7 cm.

*E. tataricum* ontogeny is complete, it has 10 ontogenetic stages (4 periods). Particulation takes place in old generative, more often in old vegetative ontogenetic stage, as a rule it does not cause regeneration, parts of taproot plant fade the same year. *E. tataricum* reproduction is performed only by seeds.

A stable existence of *E. tataricum* CP in constantly changing environmental conditions provides biodiversity of its ontogeny.

Morphological biodiversity is observed when there is a controlled burning of steppe vegetation. Rhizomatous taproot system develops. Rhizoma reaches the length no more than 3-4 cm with orthotropic direction of pigeony origin. Besides, unfavorable conditions (a controlled burning of steppe vegetation in spring and in autumn, absence of rain for a long time) cause difference in pace of development. Species does not reach mature generative ontogenetic stage and consequently graduate from young stage into old generative ontogenetic stage. Shorter duration of one of the ontogenetic stages and quick transition into the succeeding one record the acceleration of species ontogeny.

All CP under study are characterized as normal. In CP 1 a left-sided ontogenetic spectrum with absolute maximum on young generative species is formed (36,1%). In CP 2 there is a maximum on the mature generative species (35,7%), besides, there is a high percentage of species in immature (28,6%) and young generative (21,4%) ontogenetic condition.

Left-sided ontogenetic spectrum is formed due to the concentration of specimens of seed origin. The spectrum is dynamic, the number of young specimens is unstable, and they undergo a process of big elimination. Fire influences them negatively. However, a controlled burning of steppe vegetation reduces projective cover of herb layer which causes the emergence of *E. tataricum* seedlings on open territories. Germinated seeds and juvenile, immature and virgin plants fade, if there is no rain in summer period. A part of pre-generative specimens fade because of low temperature in winter period.

Bimodal ontogenetic spectrum of CP 2 proves its ability to self-maintenance with the help of seeds. The difference between young and mature generative species is insignificant which allows considering the spectrum as a left-sided variant.

In CP under study shoots and species of senile and sultrily ontogenetic condition are often absent. Absence of shoots is connected with the fact that seeds of *E. tataricum* are easily washed off by rain and blown by winds. Absence of post-generative stage are caused by quick death of old species and are typical of plants having caudex.

According to the "delta-omega" classification of L.A. Zivotovskiy (2001) both CP grow ripe. Efficiency index does not reach 0,34, age index – 0,73.

Distribution of *E. tataricum* species is accidental. It is determined by seed reproduction and peculiarities of the micro-relief (as a rule it is represented by slopes and vegetation is blown by winds). Species are distributed in groups when seeds fall from the case near female parent. It is rarely described.

## 4 Conclusion

*E. tataricum* is characterized by narrow ecological amplitude and grows exclusively in petrophytic true steppe on the territory of Khakasia. Diversity of ontogeny is found out during controlled burning of steppe vegetation and with the absence of rain. As a result of adaptation species develop rhizomatous taproot system, duration of ontogenetic conditions become shorter, possibility of seed sprouting increases because of the reduction of total projective cover of herb layer. Cenopopulation is normal and incomplete with left-sided and bimodal ontogenetic spectra. Distribution of species is accidental, determined by seed reproduction.

A research is done with financial support of the Russian Foundation for Basic Research № 18-44-190007 p_a; № 18-44-190006 p_a.

## References

1. *Red List of Threatened Species of Russian Federation. Plants and mushrooms* (M., 2008)
2. E.S. Ankipovich, D.N. Shaulo, N.V. Sedelnikova et al., *Red List of Threatened Species of Khakasia: rare and endangered species of plants and mushrooms*. (Novosibirsk, 2012).
3. A.A. Korchagin, *Field geobotany*, **3** (M.-L., Nauka, 1964)
4. T. A. Rabotnov, Tr. BIN USSR, **3**, 6 (1950)

5. A.A. Uranov, Biol. sciences, **2** (1975)
6. *Plant coenopopulations: general terms and structure* (M, Nauka, 1976)
7. *Plant coenopopulations: basic concepts and structure* (M, Nauka, 1988)
8. L.B. Zaugolnova, *Structure population of seed plants and related monitoring problems*, Diss. (St. Petersburg, 1994)
9. L.A. Zhukova, *Population Life of the Meadow Plants* (Yoshkar-Ola, Lanar, 1995)
10. L.A. Zhivotovsky, Rus. J. Ecol., 1 (2001)
11. Yu. Odum, *Ecology*, **2** (M, Nauka, 1986)

# Ecological and cenotic characteristics of communities involving the species of *Prunella vulgaris* L. and *Prunella grandiflora* L. in the Middle and Southern Urals

*Elizaveta* Bolotnik*

Russian Academy of Science, Ural Branch: Institute Botanic Garden, 620144, 8 March street, 202a, Yekaterinburg, Russia

**Abstract.** On the territory of the Middle and Southern Urals, *P. vulgaris* is found in pine, spruce-pine, birch forests, meadows and in artificial spruce-larch forest plantations; *P. grandiflora* - in birch forests, less often on meadows. *P. vulgaris* is confined to half-open places or light-forest with moistening from sub-forest-meadow to wet forest-meadow with not very rich or rather rich soils. *P. grandiflora* grows in semi-open spaces and in light forests with moistening from wet-steppe to dry-forest-meadow with rather poor soils. Both species are found on very poor soils and the ones that are poor in nitrogen, having a weak acidic and acidic soil medium. According to the soil acidity factor, the amplitude of the ecological space of *P. grandiflora* coenopopulations on studied territory goes beyond the ranges of the ecological range according to the scales of D.N. Tsyganov.

*Prunella vulgaris* L. and *Prunella grandiflora* (L.) Scholler are promising medicinal plants for the production of biologically active substances and creation of medical products. The literature mainly has information on the biology and ecology of *P. vulgaris* [1–3], while the location and phytocenotic data of the closely related species *P. grandiflora* in the natural environment of the Middle and Southern Urals require clarification. The purpose of this work was to identify typical habitats and phytocenotic confinedness of *P. vulgaris* and *P. grandiflora* in the Middle and Southern Urals, to carry out an ecological-cenotic analysis of communities featuring the studied species.

During the expeditionary research in the period from 2011 to 2014 in the Middle and Southern Urals, in various parts of the area, 20 coenopopulation groups of the genus *Prunella* were studied. The characteristics of the data collection sites of the studied samples are given in Table 1.

The description of the vegetation was carried out in the studied areas at the sites of 10x10 m [4]. The abundance of species in phytocenoses was indicated as a percentage. In our work, the scales of D.N. Tsyganov [5] were used to assess the following environmental factors in the vegetation habitats of *P. vulgaris* and *P. grandiflora* species: soil moisture

* Corresponding author: lizavb@yandex.ru

(Hd), soil trophicity (Tr), available nitrogen content in soil (Nt), soil acidity (Rc), and habitat shade (Lc). The assessment of the environmental factor in points was calculated by the generally accepted method [6–7].

**Table 1.** Characteristics of the data collection sites of the studied samples of *P. vulgaris* and *P. grandiflora*

| № | Habitat | Coordinates |
|---|---|---|
| | *P. vulgaris* | |
| **CP 1** | Aleksandrovsky sopka, Sverdlovsk region, Krasnoufimsky district, the southern slope of the hills | N 56°43'14.8"; E 057°47'08.5" |
| **CP 2** | Mokraya, Sverdlovsk region, Krasnoufimsky district, to the north of the Mariisky village - Ust-Mash, the central part of the southeastern slope of Mokraya mountain | N 56°19'51"; E 057°53'56.6" |
| **CP 3** | Yenapayevo, Perm region, near the Iren River and the Yanybaevo village, southeast slope of the mountain | N 56°34'36.5"; E 057°55'18.5" |
| **CP 4** | Sverdlovskoye, Sverdlovsk region, southeast of Krasnoufimsky district, birch forest on the southwestern slope of the mountain, northeast of the village Sverdlovskoye | N 56°11'47.8"; E 058°22'50.2" |
| **CP 5** | Sredny Munchug, the Republic of Bashkortostan, Mechetlinsky district, near the Bolsheustikinskoe village, birch forest on the southwestern slope of Sredny Munchug mountain | N 55°54'38.3"; E 058°26'26.5" |
| **CP 6** | Ilek, Chelyabinsk region, Ashinsky district, birch forest near the Ilek village | N 55°13′44″; E 58°0′5″ |
| **CP 7** | Kutushevo, the Republic of Bashkortostan, Mechetlinsky district, south-eastern slope of a mountain near the Kutushevo village, edge of birch forest | N 55°20'56.7"; E 058°29'11.1" |
| | *P. grandiflora* | |
| **CP 1** | Kurgatovo, the Republic of Bashkortostan, Mechetlinsky district, to the west of the Kurgatovo village, the right bank of the Oka river | N 56°11'39.8"; E 058°22'20.2" |
| **CP 2** | Bolshiye Klyuchi, Perm region, the northern slope, near the Bolshiye Klyuchi village | N 57°01'05.7"; E 057°025'40.1" |
| **CP3** | Ust-Utka (the Chusovaya river), Sverdlovsk region, Nature Park "Chusovaya River", the Ust-Utka village | N 57°37'51.1"; E 059°03'05.5" |
| **CP 4** | Nizhny Irginsk, Sverdlovsk region, northwestern part of Krasnoufimsky district, near the Nizhny Irginsk village | N 56°52'33.1"; E 057°27'00.5" |
| **CP 5** | Khrustalnaya, Sverdlovsk region, Pervouralsky district, pine forest near the station Khrustalnaya | N 56°50'58.6"; E 060°10'13.8" |
| **CP 6** | Sysert (the UrFU Biological Station), Sverdlovsk region, Sysert district, pine forest near the Ural State University biological station | N 56°36'4"; E 61°3'25" |
| **CP 7** | Orlovka, Chelyabinsk region, Ust-Katavsky district, pine forest near the village Orlovka, the bank of the Katav River | N 54°52′17″; E 58°5′22″ |
| **CP 8** | Kiselevo, Perm region, a meadow near the Kiselevo village | N 57°07'02.6"; E 057°20'28.0" |
| **CP 9** | Baikalovo (Shalamy), Sverdlovsk region, Baikalovsky district, birch forest near the Shalamy village | N 57°20'50.2"; E 063°39'45.6" |
| **CP 10** | Guselnikovo, Perm region, a meadow near the Guselnikovo village | N 57°19'39.71''; E 57°24'3.65'' |
| **CP 11** | Severka, Sverdlovsk region, birch forest near the Severka village | N 56°52'36.5"; E 060°18'41.8" |
| **CP 12** | Itkul, Chelyabinsk region, birch forest near the Itkul lake | N 56°9'24.59''; E 060°34'11.32'' |
| **CP 13** | Altynnoye, Perm region, spruce-larch planting near the village Altynnoye | N 56°48'25.5"; E 057°09'02.5" |

Based on 20 descriptions (7 of which with *P. grandiflora*, 13 with *P. vulgaris*), 2 dendrograms of the similarity of the communities under review for each of the studied species were constructed.

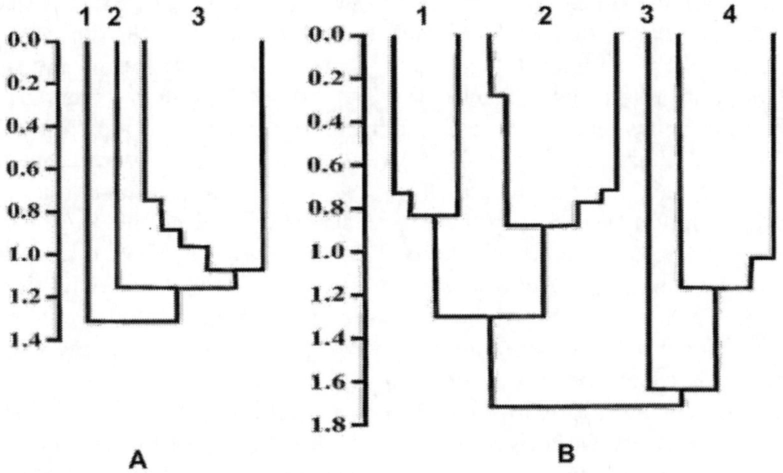


**Fig. 1.** A) Similarity dendrogram of community types involving *P. grandiflora* 1-3 – number of clusters (community types) B) Similarity dendrogram of community types involving *P. vulgaris* 1–4 – number of clusters (community types). Ward's clustering method [8], Sokal / Sneath coefficient No 4 [9].

The dendrogram (Fig. 1A) of communities' types involving *P. grandiflora* is divided into two branches at a clustering level 1, 3. The 1st cluster is represented by the description set in Krasnoufimsky district of the Sverdlovsk region, Mount Mokraya. The community is represented by a forest meadow with single trees *Betula pendula*. There are 54 species on the area of 100 m². The herb layer is dominated by: *Prunella grandiflora* (with an average projective cover of 15 %), *Trifolium medium* (10 %), *Inula hirta* (8 %) and *Gentiana cruciata* (8 %). Codominants are *Fragaria vesca* (4 %), *Origanum vulgare* (3 %), *Galium boreale* (3 %), *Fragaria viridis* (3 %), *Centaurea scabiosa* (3 %), *Knautia arvensis* (3 %), *Amoria montana* (3 %) and *Galium mollugo* (3 %).

At the clustering level of 1,2, one description (the 2nd cluster) represented by a closed birch forest, separated from the second branch. The community is described 2 km northeast of the Sverdlovskoye village, Krasnoufimsky district of Sverdlovsk region, the southwestern slope of the hill (steepness of slope 30°). The tree layer is formed by *Betula pendula* (projective cover of 45%) with a slight admixture of *Populus tremula*. Herb layer is well developed (projective cover of 90%), 60–70 cm high, has 44 species per 100 m². Herb layer is dominated by *Rubus saxatilis* (10 %), *Festuca pratensis* (8 %), *Calamagrostis epigeios* (7 %), *Brachypodium pinnatum* (7 %). The projective cover of *P. grandiflora* is 3%. The species at this point is located on the border of its range and forms a local shrinking population.

The 3rd cluster combines sparse birch forests (*Betula pendula*), bordering meadow edge communities. Tree layer projective cover is up to 30%. The height of the main canopy is 12-14 m with trunk diameter of 25-30 cm with a slight admixture of aspen (*Populus tremula*). The cenoses range from 44 to 80 species per 100 m². Herb layer is well developed (projective cover of 80-95%), 60-70 cm high. The herb layer is dominated by *Achillea millefolium* (with an average coating of 3%), *Calamagrostis epigeios* (2%), *Cerasus fruticosa* (2 %), *Galium verum* (3 %), *G. boreale* (3 %), *Filipendula vulgaris* (5 %), *Inula hirta* (2 %), *Origanum vulgare* (3 %), *Prunella grandiflora* (9 %), *Rubus saxatilis* (2 %), *Sanguisorba officinalis* (2 %), *Seseli libanotis* (4 %), *Stachys officinalis* (2 %),

*Polygonatum odoratum* (2 %), *Vicia cracca* (2 %). The projective cover of *P. grandiflora* varies from 4 to 12%. The abundance of *P. grandiflora* in these habitats cannot be called the highest, but the habitats are the most characteristic for this species.

The dendrogram of community types involving *P. vulgaris* (Fig. 1B) divided into two branches at the clustering level of 1,7. The first branch united birch forests and meadow communities near the birch forests; the second – pine, spruce-pine forests and spruce-larch forest plantation. Each of the two branches was divided into 2 clusters (community types).

The 1st cluster unites birch forest communities. The tree layer is represented by *Betula pubescens* with a slight admixture of *Populus tremula* and *Sorbus aucuparia*. The projective cover of the tree layer is from 30 to 40%, 10-12 m high, with a trunk diameter of 15-25 cm. The shrub layer is poorly developed and is formed by the *Salix caprea*. Herb layer is dense (80-85%). The dominant is *Prunella vulgaris* (7-8%), codominated by *Deschampsia cespitosa* (3 %), *Festuca pratensis* (3 %), *Fragaria viridis* (3 %), *Origanum vulgare* (3 %), *Phleum pratense* (5 %), *Pimpinella saxifraga* (3 %), *Stachys officinalis* (2 %), *Trifolium medium* (4 %), *Vaccinium myrtillus* (2 %). The relief of the studied area is gently sloping with average heights of 400-500 m above sea level.

The 2nd cluster is represented by meadow communities on the edges of birch forests. They are characterized by the inclusion of single trees *Betula pendula*, *Betula pubescens*, *Sorbus aucuparia* and *Populus tremula*. The cenoses are between 45 and 53 species per 100 $m^2$. Communities are with projective cover of up to 85% of the grass layer. *Prunella vulgaris* (7-8%) dominates, sodominated by *Bromopsis inermis* (3 %), *Dactylis glomerata* (4 %), *Festuca pratensis* (2 %), *Filipendula vulgaris* (2 %), *Fragaria viridis* (3 %), *Leucanthemum vulgare* (3 %), *Phleum pratense* (5 %), *Trifolium medium* (3 %).

The 3rd cluster is represented by the description from Perm region, near the Altynnoye village - spruce-larch forest plantation mixed with *Sorbus aucuparia* and *Padus avium*. The tree layer density is 50%. The projective cover of the grass layer is 60%. 41 species have been marked at the area of 100 $m^2$. The herb layer is dominated by *Artemisia vulgaris* (5 %), *Chamaenerion angustifolium* (7 %), *Dryopteris carthusiana* (5 %), *Dryopteris filix-mas* (5 %), *Fragaria viridis* (5 %). Projective cover of *Prunella vulgaris* is 5%.

The 4th cluster combines pine and spruce-pine forests. The cenoses range from 64 to 77 species per 100 $m^2$. The density of the tree layer is from 45 to 50%, the projective cover of the herb layer is from 75 to 90%. *Pinus sylvestris* (30–50%), 12–14 m high, with trunk diameter of 15–25 cm, prevails in the tree layer. *Picea abies* (3–20%), *Betula pendula* (1–5%) sodominate, *Padus avium* and *Sorbus aucuparia* are singled out. *Prunella vulgaris* (7–12%) is dominant in the herb layer; sodominated by *Agrostis tenuis* (3 %), *Calamagrostis arundinacea* (5 %), *C. epigeios* (2 %), *Galium boreale* (4 %), *Geranium sylvaticum* (2 %), *Rubus saxatilis* (4 %), *Vaccinium myrtillus* (4 %).

In the studied areas *P. grandiflora* is confined to semi-open spaces and light forests (from 2,9 to 3,6 points) with moisture from wet-steppe to dry-forest-meadow (from 9,0 to 10,8 points), it grows on very poor soils and soils poor in nitrogen (from 3,7 to 4,5 points), not rich (from 5,4 to 6,3 points), having acidic and weak acid medium (from 6,0 to 6,7 points) (Table 2.).

**Table 2.** Comparative ecological characteristics of habitats of *P. vulgaris, P. grandiflora*

| Ecological factors* | Amplitude of the ecological space of the studied coenopopulations | Points range of *Prunella** | Tsyganov indicator values* |
|---|---|---|---|
| | *P. vulgaris* | | |
| Hd | 9,9 - 12,8 | 7-19 | 1 - 23 |
| Nt | 4,2 - 5,1 | 1 - 9 | 1 - 11 |
| Rc | 5,8 - 6,7 | 3 - 11 | 1 - 13 |

| | | | |
|---|---|---|---|
| Lc | 2,6 - 4,3 | 1 - 6 | 1 - 9 |
| Tr | 5,4 - 6,6 | 1 - 11 | 1 - 19 |
| | *P. grandiflora* | | |
| Hd | 9,0 - 10,8 | 5 - 13 | 1 - 23 |
| Nt | 3,7 - 4,5 | 1 - 7 | 1 - 11 |
| Rc | 6,0 - 6,7 | 7 - 11 | 1 - 13 |
| Lc | 2,9 - 3,6 | 1-6 | 1 - 9 |
| Tr | 5,4 - 6,3 | 5 - 7 | 1 - 19 |

**Note**: * According to Tsyganov's scale [5]

In the studied phytocenoses, the species *P. vulgaris* is confined to half-open places or light forest (from 2,6 to 4,3) with moistening from sub-forest-meadow to moist-forest-meadow (from 9,9 to 12,8), according to the trophic scale grows on poor soils or rather rich ones (from 5,4 to 6,6), very poor and poor in nitrogen (from 4,2 to 5,1), that have an acidic and weak acid medium (pH from 5,8 to 6,7).

The data we obtained for *P. vulgaris* and *P. grandiflora* species according to ecological scales allow us to compare with those previously established by D.N. Tsyganov [5] for the European part of Russia and the Urals. It turned out that the species studied within the Middle and Southern Urals are characterized by a relatively narrow range of ecological space according to the considered ecological scales. For most factors, the amplitude of the ecological space of the studied species of the genus *Prunella* in the territory covered occupies the middle part in the amplitude of the ecological space of the species and does not go beyond the ecological range of species according to the scales of D.N. Tsyganov [5].

In terms of soil conditions, the *P. vulgaris* species reaches its potential to its fullest in relation to soil moistening factor and less in the richness of nitrogen in the soil. *P. grandiflora* for the totality of soil conditions most fully realizes its potential in relation to the factor of soil moistening and less - in acidity. At the same time, according to the acidity of the soil, the species occupies an extreme position - 6-6,7 points (7-11 on D.N. Tsyganov scale) and goes beyond the ranges of the ecological area on the scales of D.N. Tsyganov, which is probably due to the presence of *P. grandiflora* on the northern boundary of the range. *P. vulgaris* in the Middle and Southern Urals grows in habitats from well shaded to poorly lighted ones, on soils with sufficient moisture and nitrogen content. While *P. grandiflora* prefers drier and lighter habitats, with less rich soils and low nitrogen content. On the territory of the Middle and Southern Urals, *P. vulgaris* is found in pine (*Pinus sylvestris*), spruce-pine (*Pinus sylvestris*), birch (*Betula pubescens*) forests, in meadows and in artificial spruce-larch plantations. *P. grandiflora* coenopopulations are found in birch forests (*Betula pendula*), and are less common in meadows, which is probably related to the position of this species on the northeastern boundary of its range.

## References

1. L.I. Abramova, Biological flora of the Moscow region, **12** (1996)
2. I.N. Barsukova, Grow. resources., **3** (2014)
3. E.S. Zakamskaya, *Ontogenetic atlas of medical plants*, **4** (Yoshkar-Ola, 2004).
4. B.M. Mirkin, *Phytocenology: principles and methods* (M., Nauka, 1978)
5. D.N. Tsyganov, *Phytoindication of ecological regimes in subzone of Coniferous-Broadleaved forests* (M., Nauka, 1983)

6. E.V. Zubkova, L.G. Khanina, T.I. Grokhlina, Y.A. Dorogova, *Digital processing of geobotanical descriptions using ecological scales and EcoScaleWin Software: manual* (Yoshkar-Ola, Mari State University, 2008)
7. P.S.Shirokikh, Agrar. Russia, 1a (2009)
8. J.H. Ward, J. of the Amer. Statistical Association, **58**, 301 (1963)
9. R.R. Sokal, P.H.A.Sneath, *Principles of numerical taxonomy* (San Francisco-London, W.H. Freeman and Company, 1963)

BIO Web of Conferences **16**, 00006 (2019) https://doi.org/10.1051/bioconf/20191600006
*Results and Prospects of Geobotanical Research in Siberia*

# Rare species of psammophyte flora in transboundary areas of Southern Siberia and Mongolia

*Vladimir* Doronkin[1*], *Natalya* Vlasova[1], *Nanjidsuren* Ochgerel[2], *Luvsanbaldan* Enkhtuya[2] and *Tovuudorj* Munkh-Erdene[2]

[1] Central Siberian Botanical Garden, Siberian Branch of the Russian Academy of Sciences, Zolotodolinskaya st., 101, 630090, Novosibirsk, Russian Federation
[2] Institute of General and Experimental Biology, Mongolian Academy of Sciences, Jukov st., 77, 210351, Ulaanbaatar, Mongolia

**Abstract.** Data on distribution of 3 rare species which occur in sandy arrays of cross-border territories of Southern Siberia and Mongolia are provided: *Iris psammocola*, *Iris loczyi* (Iridaceae) and *Vicia tsydenii* (Fabaceae), a brief characteristic of their cenopopulation is given.

The originality of the plant communities which occur in sandy arrays allows a number of authors to distinguish them as special florocenotype-psammophyton (Psammophyton) [1]. The structure of vegetative cover of sandy arrays considerably depends on the relief and degree of the substrate coherence [2]. In the south of the Tuva Republic in the reserved area "Ubsu-Nur Depression" the cluster of "Tsuger-Els" was investigated. It represents a sandy mane extended from the northwest to the southeast along the left riverbank of the Tes-Hem. To the northwest of mane there is big fresh lake Tore-Hol. Within the cluster a wavy undulating land prevails. The total area of Tsuger-Els sands covers about 60 km$^2$. Similar arrays proceed also in the territory of Mongolia - the Altan-Els sands entering the largest array Borig-Del. The cluster belongs to Ubsunursky plain desert-steppe province [3]. In 2005 in the territory of Tsuge-Els cluster a new species for the flora of Russia - *Iris psammocola* Y.T. Zhao was found [4]. The species has obconical rhizome, flowers on short peduncles which are inside the curtain is characteristic of this species, is on 2 pedicel lanceolate webbed spata are available for each flower, the perianth tube is thin funnel-shaped, many times exceeds the ovary at length, the flowers are sulfur-yellow. In 2013 in the neighborhood of lake Tore-Hol we also found this species: "Republic of Tuva, Erzinsky district, lake Tore-Hol neighborhood, Tsuger-Els sands, N 50°09.231 'E 95°12.751', H=1200 m, 6.06.2013, V. Doronkin, N. Vlasova" (NSK). The plants occur in different plant aggregations of small structure and the rarefied herbage, at the same time the projective covering makes about 20% in cenopopulation: *Thymus mongolicus* (Ronn.) Ronn. - sp., *Carex sabulosa* Turcz. ex Kunth - sol., *Agropyron desertorum* (Fisch. ex Link) Schult. - sp., *Thesium tuvense* Krasnob. - sol., *I. psammocola* - sol.; in some about 30%: *Alissum lenense* Adams - sol., *Th. mongolicus* - sol., *Artemisia sp.* - sol., *Oxytropis tragacanthoides*

* Corresponding author: norbo@ngs.ru

Fisch. - sp., *Gypsophila patrinii* Ser. - sp., *Orostachys spinosa* (L.) C.A. Mey. - sol., *I. psammocola* - sp. The fullest data on vegetation of these areas are provided in the book "Steppes of Inner Asia" [5].

For the first time *I. psammocola* in Mongolia was found by us in 2013 at inspection of the adjacent territory: 1) "Mongolia, aimag Uvs, Altan-Els sands, N 49°46 'E 95°13', H=1370 m, sand hills, 9.06.2013, L. Enkhtuya, T. Munkh-Erdene, V. Doronkin, N. Vlasova", 2) "sands of Borig-Del, N 49°50.26' E 94°20.53', H=1117 m, sandy barkhans, 11.06.2013, the same authors" (UBA, NSK) [6]. On sandy arrays of Altan-Els the projective covering in tsenopopulyation reached 30%: *I. psommocola* - sol., *Th. mongolicus* - sol., *Potentilla acaulis* L. - cop.□, *Carex duriuscula* C.A. Mey. - sp., *Thesium refractum* C.A. Mey. - sol., *Artemisia frigida* Willd. - sol., *Koeleria altaica* (Domin) Kryl. - sol., *Agropyron cristatum* (L.) Beauv. - sol. *I. psommocola* occurs only in sandy arrays: in Russia - the South of Tuva (Tsuger-Els sands), the South of Buryatia (surrounding Kyakhta), in Mongolia - the aimag Uvs (Altan-Els sands, Borig-Del sands), in China - the Tenger sands relating to the Alashan Desert. Thus, the species has disjunctive area in Central Asia.

Another rare species – *Iris loczyi* Kanitz also occurs only in the south of the Tuva Republic. The species has leaves greyish-green glaucous, rigid, twisting, the 1-3rd., flowers single blue-violet, nimbus tube up to 14 cm length. Its locations are known around lake Tore-Hol, the mountain Tsagan-Tologoy and Anchalau. It is found by us on border with Mongolia: "Republic of Tuva, Erzinsky district, valley of river Tes -Hem, N 49°59' E 95°30', H=1280 m, sandy arrays, 7.06.2013, V. Doronkin" (NSK).

The projective covering in some cenopopulation reaches from 20%: *I. loczyi* - sol., *Caragana bungei* Ledeb. - sol., *Carex duriuscula* - sol., *Koeleria cristata* (L.) Pers. s.l. - sp., *Festuca valesiaca* Gaudin - sp., up to 30%: *I. loczyi* - sol., *Festuca valesiaca* - cop.□, *A. frigida* - sp., *P. acaulis* - sol. The species in Mongolia was known only from Mongolian Altai [7], it was found by us also in the Botaniko-geographic region - Depression of Great Lakes: 1) "Mongolia, aimag Uvs, vicinities of the lake Bayan-Nour, N 49°56.790' E 94°04.522', H=940 m, Borig-Del sands, sandy barkhans, 10.06.2013, L. Enkhtuya, T. Munkh-Erdene, V. Doronkin, N. Vlasova, ", 2) in the same place, at Lake Boyan-nur, N 50°01.36' E 94°01.20', H=994 m, sands, 11.06.2013, the same authors " (UBA, NSK). The projective covering was from 20%: *I. loczyi* - sol., *Chamaerhodos sabulosa* Bunge - sol., *P. acaulis* - sol., up to 30%: *I. loczyi* - sol., *Ch. sabulosa* - sp., *Ephedra monosperma* C.A. Mey. - sol., *P. acaulis* - sol., *Astragalus brevifolius* Ledeb. - sol., *C. duriuscula* - sol. Also the species is found in the Southwest part of Altan-Els sands: "Aimag Uvs, southwest part of Altan-Els sands, N 49°46.857' E 95°06.049', H=1200 m, hilly sands, 12.06.2013, the same authors" (UBA, NSK). The projective covering made from 35%: *I. loczyi* - sol., *F. valesiaca* – cop.□, *Artemisia obtusiloba* Ledeb. - sol., *Hypecoum erectum* L. - sol., up to 30%: *I. loczyi* - sol., *Hypecoum erectum* - sp., *F. valesiaca* - sol., *Allium ramosum* L. - sol. The species is widespread in Central Asia, in Southwest Asia, Mongolia, China. Locations in Tuva are the most northern.

Thus our researches confirm the data that the psammophytic complex of flora of Tuva is directly connected with the Mongolian flora which is explained by cross-border located array of Tsuger-Els [8].

Another area of psammophytes distribution - the South of the Buryatia Republic: *Vicia tsydenii* Malyschev is a rare species here. Low perennial approx. 20 cm of height, with thin long rhizome. The stalks are direct, branchy. The leaflets among 4-5 couples, oblong and obovate at 10-15 mm length, from below greyish due to the pressed beard of hairs. Flowers of 5-7 mm length, pinkish-violet, in friable brushes, beans wide with 1-2 seeds.

The species is described from the neighborhood of the village Kiran, Kyakhta region of the Buryatia Republic. Later it was found also on sand hills 26 km away along Kieran-

Thamir Road. In Mongolia it is known only from the Mongolo-Daursky botaniko-geographic region, occurs on the Selenga River, the Orkhon River, in the neighborhood of the settlement Shamar (UBA, MW, LE). We also found it in the valley of the Orkhon River: "Mongolia, aimag Selenge, Sum Shaamar, in the valley of the Orkhon River, N 50°06.675' E 106°11.584', H=620 m, on sand hills, 24.06.2012, N. Vlasova, V. Doronkin " (NSK). It has been found in low-specific community, projective covering 20%: *V. tsydenii* - sol., *Leymus littoralis* (Griseb.) Peschkova - sp., *Thymus baicalensis* Serg. - sp., *Asparagus gibbus* Bunge - sol. The species is concentrated only in the basin of the Selenga River, which flows from the South to the North across the territory of Mongolia and Southern Siberia.

**Fig 1.** Altan-Els sands

**Fig. 2. A)** *Iris psammocola* Y.T. Zhao **B)** *Iris loczyi* Kanitz **C)** *Vicia tsydenii* Malyschev

## References

1. R.V. Kamelin, Bot. Zhurn., **72** (1987)
2. N.I. Akzhygitova, S.-W. Breckle, G. Winkler, E. A. Volkova, W. Wucherer, L. Ja. Kurochkina, G. B. Makulbekova, N.P. Ogar, E.I. Rachkovskaya, I.N. Safronova, V.N. Khramtsov, *Botanical geography of Kazakhstan and Middle Asia (desert region)* (St. Petersburg, 2003)
3. V.M. Hanminchun, N.V. Sedel'nikova, N.V. Perova, *Flora Cuger-Jeliss Ubsunurskoj kotloviny* (Barnaul, 1997)
4. V.M. Doronkin, D.N. Shaulo, Bot. J., **92**, 435 (2007)
5. I.M. Gadghiev, A.Yu. Korolyuk, A.A. Tytlyanova et al., *Steppes of Inner Asia* (SB RAS Publisher, Novosibirsk, 2002)
6. L. Enkhtuya, T. Munkh-Erdene, Sci. Proc. Inst. Bot. Mongolia, **25**, 174 (2013)
7. M. Urgamal, B. Oyuntsetseg, D. Nyambayar, Ch. Dulamsuren, *Conspectus of the vascular plants of Mongolia* ("Admon Printing" Press., Ulaanbaatar, 2014)
8. N.A. Dulepova, A.Yu. Korolyuk, Rast. Mir Az. Rossii, **2** (**10**) (2012)

# Rare Plants and Plant Communities of the Projected "Togul" National Park (Altai Region)

*Natalja* Elesova, *Marina* Silanteva* and *Liudmila* Sokolova

Altai State University, Biological Department, Chair of Botany, 656049 Prospect Lenina, 61, Barnaul, Russian Federation

**Abstract.** The value of the Salair Ridge territory for the creation of the "Togul" National Natural Park is determined primarily by the natural complex of black coniferous forests with the participation of relict plant species of the Miocene-Pliocene coniferous-deciduous forests and a group of early flowering ephemeroid plants, old-aged *Pinus sibirica*, *Picea obovata* and linden forests and a variety of tall grasses. Based on the performed geobotanical descriptions, the phytocenotic characteristic of linden forests was compiled. Information is provided on the location of 10 rare and endangered species of plants and fungi found during a survey of the territory of the planned national park.

## Introduction

One of the most effective measures for the conservation of natural ecosystems and biological diversity is the creation of national natural parks – specially protected natural areas, which are conservational, environmental education and research institutions. According to the law of Russian Federation 33-ФЗ dated March 14, 1995 "On specially protected natural territories", the following main tasks are assigned to national parks: a) conservation of natural complexes, unique and standard natural sites and objects; b) preservation of historical and cultural objects; c) environmental education of the population; d) creating conditions for regulated tourism and recreation [1-4].

The creation of a protected natural area of federal significance is planned on the Salair Ridge - in three districts of the Altai Region along the border with the Kemerovskaya Oblast (Fig. 1). The goal of creating the "Togul" National Natural Park is to preserve the typical and unique ecosystems of the taiga low-hill terrains of the Salair Ridge. The decision on its creation was made at the meeting of the working group at the Ministry of Natural Resources and Ecology of Altai Region within the framework of the execution of the May decree of the President of the Russian Federation "On the national goals and strategic objectives of the development of the Russian Federation for the period up to 2024". The park area will be about 160240 hectares. It will include the territories of three districts: Zarinsky – 24130 hectares (4.7% of the district area); Togulsky – 77745 hectares (38.5 %) and Eltsovsky – 58365 hectares (26.9 %). It is also planned to include into the

* Corresponding author: msilan@mail.ru

borders of the national park two existing reserves of regional significance – "Togulsky" and "Eltsovsky" [5].

In 2015-2016 employees of the Gebler Ecological Society and the Chair of Botany of the Altay State University conducted a survey of the vegetation cover of the projected "Togul" national park. The collected herbarium in the amount of 500 herbarium sheets and geobotanical descriptions made it possible to identify the locations of 10 rare and protected plant species (all of them are listed in the Red Book of Altai Region, two of them are listed in the Red Book of the Russian Federation) and characterize linden forests communities.

## Botanical and geographical characteristics of the territory

The territory of the projected "Togul" National Park is located within the western slope and the southern part of the Salair Ridge. In appearance, the Salair Ridge resembles a hilly plain. The highest point is Mount Kivda (621 m). The whole area is indented by a highly branched river network, gullies and ravines. The forests of Salair favor the formation of precipitations and the aggravation of frontal processes, especially when moving south. Therefore, the territory is characterized by a special climate, which Slyadnev [6] called the "Salair forest climate". The continental climate here is slightly reduced in comparison with the surrounding area: summers are cooler and winters are warmer. The average temperatures in January are -16...-17°C, and in July +16...+18°C. Precipitations ranges from 500 mm in the north to 800 mm in the south. In winter, a thick snow cover of more than a meter high is formed [6].

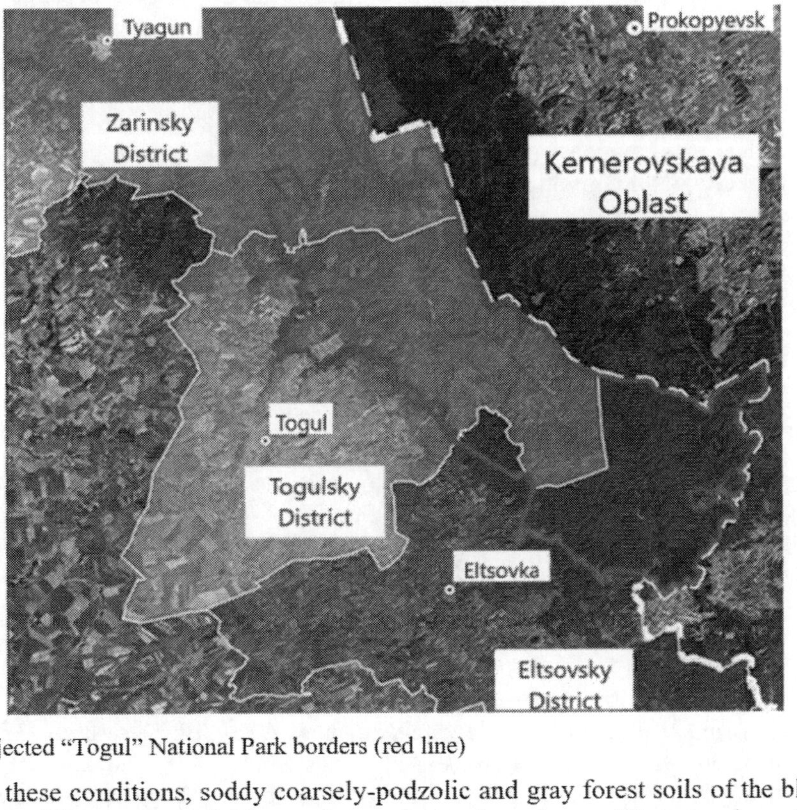


**Fig. 1.** Projected "Togul" National Park borders (red line)

Under these conditions, soddy coarsely-podzolic and gray forest soils of the black taiga of the Salair ridge are formed. The mother rocks are loess loams. They are characterized by a high humus content reaching in the upper horizon according to P. I. Shavrygin and A. N.

Rozanov [7] 3.6–6.0 %, and according to L.M. Burlakova, V.A. Rassypnov end L.M. Tatarintsev 7-8 % [7, 8].

The territory of the projected "Togul" National Park is known primarily for its black coniferous forests, which are the custodians of relict plant species of the Miocene-Pliocene coniferous-deciduous forests: *Tilia sibirica* L., *Brunnera sibirica* Stev., *Dentaria sibirica* (O.E. Schulz) N. Busch, *Asperula odorata* L., *Asarum europaeum* L., *Sanicula europaea* L., *Brachypodium sylvaticum* (Huds.) P. Beauv. and others [9]. As part of the black taiga, a rare community even on the scale of South Siberia, there will be plots of linden forests. Old-age *Pinus sibirica* Du Tour and *Picea obovata* Ledeb. forests preserved in hard-to-reach valleys of taiga rivers are also valuable. The age of the oldest trees here reaches several hundred years, and individual *Pinus sibirica* trees up to 800 years.

The following altitudinal belts are distinguished within the natural park on the Salair ridge: the belt of birch forests and steppificated meadows of the western slope on podzolized chernozems and dark gray forest soils (northern forest-steppe) and the belt of black forests on gray forest and deeply podsolized soils. The black taiga consists of *Abies sibirica* and in shifts of *Populus tremula*, in places there are pine, larch and birch. In the valleys of the Chumysh River confluents, one can find aspen-fir forests mixed with spruce [10].

A distinctive feature of Salair's forest phytocenoses is tall grass, which grows on very poor soils. It is developed in more open spaces and burned areas: *Aconitum excelsum* Rchb., *Cacalia hastata* L., *Euphorbia lutescens* Ledeb., *Angelica sylvestris* L., *Festuca gigantea* (L.) Vill. end others. In addition, mass spring flowering of early bloomers, including up to 10 types of ephemeroids, is a characteristic of these black forests: *Corydalis bracteata* (Steph. ex Willd.) Pers., *Anemone altaica* Fisch. ex C.A. Mey., *Erythronium sibiricum* (Fisch. et C. A. Mey.) Kryl., *Pulmonaria mollis* Wulfen ex Hornem, *Ranunculus monophyllus* Ovcz. end others.

Russian researchers call the black taiga "rain forests of temperate latitudes", as the amount of annual precipitations approaches 1000 mm. A preliminary survey of the territory revealed 43 species of animals and 29 species of plants and fungi listed in the Red Books of various ranks.

## Rare plant and fungy species list

**Polypodiaceae Bercht. Et J. Presl.**
*Polypodium sibiricum* Sipl. Rarely [11].
- Eltsovsky district, Salair ridge, the river Chumysh, outskirts of the river Tatarka. Shaded rocks near the water.
- Eltsovsky district, Salair ridge, Chumysh river, rock Bakharev stone. Moss cliffs on the bank of the Chumysh. N53°18.211' E86°23.556'.

**Aristolochiacaae Juss.**
*Asarum europaeum* L. Dark coniferous, *Abies* and linden forests. Usually [11].
- Eltsovsky district, near the village Mostovaya. Aspen-*Abies* forest. N53°45.07' E86°76.91'.

**Tiliaceae Juss.**
*Tilia sibirica* L. Dark coniferous forests. Rarely [11].
- Zarinsky district, 1 km from the river Krutishka. "Togulsky" reserve. *Matteuccia-Aegopodium-Aconitum* linden forest. N53°42.075' E85°59.353'.
- Zarinsky district, 3 km north-east of the village Togulenok. "Togulsky" reserve. Herbs-*Aconitum-Aegopodium* linden forest. N53°41.503' E85°59.306'.
- Zarinsky district, 2.2 km to the south-west from the village Togulenok. "Togulsky" reserve. *Oxalis-Aegopodium-Matteuccia* linden forest. N53°42.210' E85°59.547'.

- Zarinsky district, 2.2 km to the south-west from the village Togulenok. "Togulsky" reserve. *Festuca-Aegopodium- Asarum* aspen-linden forest. N53°42.181' E86°00.119'.

**Thymelaeaceae Juss.**

*Daphne mezereum* L. *Abies* and birch-*Abies* forests. Rarely [11].

- Eltsovsky district, near the village Mostovaya. *Pinus sibirica-Picea-Abies* forest. N53°46.05' E86°78.09'.

**Boraginaceae Juss.**

*Brunnera sibirica* Stev. Forest meadows, along rivers and streams. Rarely [11]. River valley Srednij Uksunai, to the confluence of the Kedrovka river. Floodplain forest.

**Campanulaceae Juss.**

*Campanula trachelium* L. Dark coniferous forests. Very rarely [11].

- Eltsovsky district, Salair ridge, the river Chumysh valley. Swamped forest. N53°25.022' E86°48.488'.
- Eltsovsky district, Salair ridge, the river Chumysh valley. Taiga on the slope of the Chumysh valley. N53°18.053' E86°23.175'.
- Togulsky district, the neighborhood of the village of Malaya Rusyanka. Birch with sporadic *Abies* forest. N53°34.281' E86°02.154'.
- River valley Srednij Uksunai, to the confluence of the Kedrovka river. Floodplain forest near the swamp.

**Liliaceae Juss.**

*Erythronium sibiricum* (Fisch. et C. A. Mey.) Kryl. Dark coniferous and deciduous forests, forest meadows and forest edges. Usually [11].

**Araceae Juss.**

*Calla palustris* L. Marshes, marshy shores of ponds. Rarely [11].

- Eltsovsky district, the village of Sary-Chumysh, lowland bog. N53°24.54' E86°75.73'.
- Tolgulsky district, the village Podsopka surroundings, lake Chertovo, swampy shallow water. N53°56.12' E86°10.23'.

**Lobariaceae Chevall.**

*Lobaria pulmonaria* (L.) Hoffm. Dark coniferous forests. Rarely [11].

- Zarinsky district, 2.2 km to the south-west from the village Togulenok. "Togulsky" reserve, h 299 m, northeast slope, *Oxalis-Aegopodium-Matteuccia* linden forest. N53° 42.21,0' E85° 59.54,7.

**Phallaceae Corda**

*Phallus impudicus* L. Coniferous, deciduous and mixed forests. Rarely [11].

- Zarinsky district, 3 km north-east of the village Togulenok, h 317 m. Herbs-*Aconitum-Aegopodium* linden forest. N53° 41.50,3' E85° 59.30,6'.

## Phytocenotic characteristics of linden forests

On the basis of the geobotanical descriptions made, a phytocenotic characteristic of the Altai Region linden forests has been compiled.

The tree stand is usually two-layered 22-23 m high formed by *Tilia sibirica*, *Populus tremula* L., *Abies sibirica*, *Betula pendula* Roth and *Betula pubescens* Ehrh. These forests are thick and wet enough; have 70-80 % canopy cover. The average age of Tilia sibirica thees is 60 years old, the average diameter of the linden stems is 23 cm, and the maximum is 38 cm. *Abies sibirica* and *Tilia sibirica* are found in the undergrowth. Common projective cover of the shrub layer is 3-10 %, dominant species are *Sorbus sibirica* Hedl., *Caragana frutex* (L.) K. Koch, *Ribes atropurpureum* C.A. Mey. With high constancy there are *Padus avium* Mill., *Lonicera xylosteum* L., *Sambucus sibirica* Nakai. There is no shrub layer sometimes.

The herbaceous layer has two or three sublayers. Its common projective cover is 65 % (maximum is 85 %). The first and the second sublayers (tall-grass) dominant species are *Crepis sibirica* L., *Matteuccia strutiopteris* (L.) Tod., *Aegopodium podagraria* L., *Aconitum septentrionale* Koelle, *Festuca altissima* All., *Urtica dioica* L., *Pleurospermum uralense* Hoffm. Small grass is presented by *Asarum europaeum*, *Oxalis acetosella* L., *Cerastium pauciflorum* Steven ex Ser. and others. There are 56 species of higher vascular plants per 400 $m^2$, including 6 species of herbaceous tertiary relics. The total population area of *Tilia sibirica* in the Altai Region is about 2 $km^2$.

## Conclusion

Thus, as a result of the study, new information was obtained on the finding of rare and endangered plant species, and the phytocenotic characterization of linden forests was given. The data obtained substantially supplement the idea of the real value of the natural complexes of the Salair Ridge for the creation of the "Togul" National Natural Park.

## References

1. M. Silantyeva, P. Kosachev, A. Kuznetzov, A. Grebennikova, Int. J. of Env. St., **71 (5)** (2014)
2. P. Kosachev, E. A. Kryuchkova end P. D., Acta Biologica Sibirica, **3 (4)** (2017)
3. S. A. Sheremetova, A. L. Ebel, T. E. Buko, Bull. of Tomsk St. Uni.: Biol., **1 (17)** (2012)
4. A.V. Sobisevich, V.A. Snytko, Bull. of Moscow St. Reg. Uni. Series: Nat. Sci., **4** (2018)
5. *Newspaper "Altai Pravda"* http://ap22.ru/paper/Pervyy-natsional-nyy-park-Togul-poyavitsya-v-Altayskom-krae.html (2019)
6. A. P. Slyadnev, *Sketches of climate of Altai region* (Barnaul, Alt. publ., 1958)
7. *Soils of Altai Region* (M., Publ. of USSR Academy of Sciences, 1959)
8. L.M. Burlakova, V.A. Rassypnov, L.M. Tatarintsev, *Soils of the Altai Territory* (Barnaul, Alt. SKHI, 1988)
9. N. V. Elesova, Probl. of Bot. of South Siberia and Mongolia (2018)
10. V.D. Alexandrova, N.P. Guricheva, L.I. Ivanina, *Natural zoning of Altai Krai* (M., Science, 1958)
11. *Red Book of Altai Region*, **1** (2016)

# Some features of rare and endangered plant species introduction

*Tatyana* Elisafenko* and *Olga* Dorogina

Central Siberian Botanical Garden of the SB RAS, Zolotodolinskaya Str., 101, Novosibirsk, 630090, Russia

**Abstract.** The article discusses the features and main stages of rare and endangered plant species introduction. It is necessary to be guided by the principles of priority and expediency in the preparatory stage, which includes the analysis of literary data and preliminary studies in nature. The first stage is the primary introduction; it involves a detailed study of natural populations and organizing an introduction site. Based on the obtained results, the acclimatization assessment is determined. The study of biological features including reproduction is carried out at the second stage. The third stage includes investigations of the introduced populations. The phenorhythm stability, limiting factors determining dates of phenophase onset, the introduction population optimal size and number, seed longevity and conditions of their storage, self-maintenance ability are revealed. As a result, it is given the adaptation assessment, and recommendations are developed to save rare and endangered species in nature and under introduction conditions. Additional studies include applying karyological, molecular-genetic, biotechnological techniques, and possibilities to reconstruct natural populations.

Since the mid XX century, a prior activity trend of introduction centers has been creating a collection or exposition of rare and endangered plant species. The International Botanic Gardens Plant Conservation Program [1] presents ways to develop own biodiversity conservation programs and tools for botanical gardens. The priority is the biodiversity conservation and rational use in the region of the introduction center location. The complete conservation of plant biodiversity involves applying different methods, and the governmental, public organizations and interested entrepreneurs cooperation.

Introducing rare and endangered plant species applies traditional techniques, however, the there are some peculiarities - it takes more time and investments, and the results are often negative. At present, it is necessary to develop guidelines for the gradual study and preservation of rare and endangered plant species in the introduction centers. Researches should be aimed to study the biological and physical processes affecting the biodiversity diminution, as well as to identify its causes, and to develop tools for the regional biodiversity conservation and rational. The introduction as a measure of conservation turns out to be inexpedient for some of these species. Otherwise, introductions may be the only way to preserve other species.

* Corresponding author: tveli@list.ru

The work objective is to develop guidelines for such species introduction related to the relevance of rare and endangered species conservation, studying their biological features, identifying useful properties.

Preparatory and primary introduction are important steps in introducing rare and endangered species. The preparatory stage includes the object selection, where the main principles are priority and expediency [2]. Besides, it is necessary to carry out a preliminary population study at this stage, because the low abundance and insignificant knowledge of this object hamper to introduce and reproduce rare and endangered species. Comprehensive expeditionary research, including specialists of various profiles (florists, geobotanists, taxonomists, biomorphologists, introductors) will provide a fruitful search for populations, valid habitat characterization, the population state description and receiving the material. The introduction of rare and endangered species populations should begin with studying the natural population: its condition assessment, including demographic indicators [3, 4], plant vitality [5], ontogenesis and reproductive characteristics. It is necessary to be guided by the program for studying the species of red-books lists.

The obtained herbarium, field and literature data are analyzed and the introduction forecast is made based on some approaches: the ecological-historical analysis by M.V. Kultiasov, the flora genetic technique of choosing introduced species by K.A. Sobolevskaya, the generic complexes method by F.N. Rusanov, the tool of accounting the acclimatization experience in the past by A.N. Avrorin, the biomorph comparison by M.T. Mazurenko, the phytocenotic one by R.A. Karpisonova. Highly specialized and stenotopic species are unpromising with high probability. However, only a direct experiment can reveal the introduction success of a particular species, as the interspecific competition elimination, natural habitats imitation, agrotechnical measures may be sufficient conditions for plant growth and development. The preparatory stage is resulted in the introductory passport, which includes the data of the natural initial population: a geographical location, habitat topography, date, collector, habitat phytocenotic and ecological description, plant community species composition (estimate the species participation in it), evaluating a population state, demographic indicators, viability plants, data on ontogenesis and reproductive characteristics study.

The first stage includes the material search and selection, assessment of the plant primary introduction (acclimatization), reproduction and study of introducent biological features. Diaspores can be obtained from nature or introduction centers. The latters may contain genotypes significantly different from natural populations. Therefore, the refinement of the selected species systematic position is mandatory, if necessary, identification is carried out by molecular-genetic methods. Particular attention is paid to the high representativeness and maintenance of the taxa genetic purity in developing the methodology of rare species reproduction and preservation *in vitro*. Therefore, one of the most important stages of rare species and genotype preservation is identifying the initial and micropropagated plants by ISSR markers. Some researchers consider that rare and endangered species introduction by seeds obtained by delectus from other regions is unacceptable. Such species introduction should be immaculately clean, using seeds from natural populations for this purpose [6]. It is also a controversial statement that rare species introduction should be carried out only by seed, because it is necessary to exclude plants removal from nature. In fact, living plant removal in limited quantities from plural populations is permissible, as well as vegetatively mobile plants. An important issues is the representativeness of the rare and endangered species population under conditions of introduction. The recommended optimal area and number of introduced populations are the following: 100 $m^2$ (5–10 individuals) for trees, 50 $m^2$ (10–30 individuals) for shrubs, 40 $m^2$ (100–500 individuals) for large grasses, 10-20 $m^2$ (500–1000 individuals) for small-shrub and non-roselatte herbs [6].

Expositions and collections of rare and endangered plant species should be organized according to an ecological principle creating sites with microenvironmental conditions for mesophytes, xerophytes, petrophytes, hygrophytes, etc. Rare and endangered plant species can be represented both introducing populations (in the presence of a large number of individuals and changes in generations), and samples consisting of several individuals. Every variant is realized as monoculture, or as a phytogroup part. Monoculture plots are usually supported by agrotechnical measures (weeding, thinning, loosening, fertilizer), this allows maintaining plants with good vitality and producing seed every year. Not all rare and endangered species can exist in monoculture conditions, it's necessary to create phytogroups (areas imitating plant communities) for some of them. It is advisable to form phytogroups for educational activities as various micro-exposures. However, dense planting of different species will increase competitiveness, allelopathy may occur, therefore, seed productivity decrease will take place. In addition, if the activity of the introduction center is carried out as part of the global task of introducing rare and endangered species to preserve species diversity, it is advisable to create and maintain hybrid populations. Hybrid populations are individuals of the same species from different points of its range. This is especially true for cross-pollinated plants of the same species grown in a small area without isolation. Hybrid populations increase the probability of successful existence under introduction conditions, and they can be used as donors to reconstruct the disappeared and small populations. During the first stage, the species systematic identity is verified, plants survival rate (the number of introducents survived of the seedlings number), winter hardiness, drought tolerance is determined. The following data should be added to the existing introduction certificate: date of introduction, number of individuals or seeds, location on the introduction site.

The second stage includes introducents' reproduction and their study. It reveals the ability to seed and vegetative reproduction; defines seed germination feature; describes ontogenesis, biological characteristics of growth and development; evaluated the primary introduction, which reflects plants acclimatization to the introduction center conditions.

The third stage is studying the introduced populations. It reveals the stability phenorhythm, limiting factors determined the dates of phenophase onset, an optimal size and number of the introduced population, seeds longevity, their storage conditions, self-maintaining ability; elaborates recommendations to preserve rare and endangered species in nature and under the conditions of introduction. The main study directions at this stage are to determine the possibility of introduced species (seed or vegetative) reproduction and adaptation assessment. The result is compiling a biological introducing passport of a species, which includes the research results of three stages: name (Latin and Russian), synonyms, categories of rarity according to different lists of the Red Books, species range and habitat (on literary sources), ecological group, presence of herbarium specimens in herbaria of Russia, species locus classicus, photos of herbarium and natural specimens, morphobiological characteristics in nature (on literary data and original research), life form, seed collecting time, collector name, morphology of fruits and seeds: length, width, weight, germ characteristics, seed reproduction features - germination conditions (light, darkness , positive high and low temperatures, their effect duration stage-by-stage), period before germination, period of germination, germination, durability, especially fruiting, percentage of fruit production, seed production, self-seeding ability, dates and methods of sowing seeds, features of ontogenesis, rhythmological group, diseases, pests of introduced plants, survival rate (the ratio of planted individuals and definitive individuals number), agrotechnical measures applied to introduced species, assessment of the species economic properties, adaptation degree of plants , recommended measures to this plant species conservation.

Creating introductory collections allows carrying on complimentary research: karyological (number of chromosomes), phytochemical (biological active substances of many rare and endangered species are not identified), molecular-genetic, biotechnological ones. The ISSR-marking method has recently been used in population and taxonomic studies, as ISSR markers have a large variability. A modified SDS electrophoresis method in polyacrylamide gel of seed storage proteins [7, 8] and a technique of genomic DNA inter-microsatellite regions with ISSR markers [9] can successfully analyze intra- and interpopulation variability and genotyping of individual samples. This study will help to identify adaptation processes occurring at the genetic level, determine the magnitude of genetic variability, control genetically the formation of rare and endemic species stable populations at different stages, and develop recommendations for their conservation. Evaluating the genotype frequency and identifying the most adapted genotypes, which are the basis for the formation of populations in different years, can be carried out on the basis of an annual analysis of electrophoretic spectra of seed storage proteins in populations growing under the introduction conditions. Biotechnological approach application offers fundamentally new opportunities to preserve and reproduce the plant gene pool. Despite the importance of maintaining genetic diversity in natural ecosystems (*in situ*), it has recently become clear that many rare species cannot be preserved without developing *ex situ* technologies. An important step is to research plant morphogenesis *in vitro* using light and scanning electron microscopy, as well as create a crop bank *in vitro,* and optimize conditions for its maintaining.

## Acknowledgements

The study is carried out according to of the state assignment of the Central Siberian Botanical Garden of the Siberian Branch of the Russian Academy of Sciences, state registration no. AAAA-A17-117012610051-5 and applying bioresource scientific "Collection of Living Plants in Open and Closed Ground" USU 440534.

## References

1. W. Jackson, L.A. Sutherland, *International Agenda for Botanic Gardens in Conservation* (Botanic Gardens Conservation International, U.K, 2000)
2. T.V. Elisafenko, O.V. Dorogina, *Problems of industrial botany in advanced industrial regions* (Kemerovo, 2018).
3. L.A. Zhukova, Journal of general biology, **44**, 3 (1983)
4. L.A. Zhivotovsky, Rus J of Ecology, 1 (2001)
5. Yu.A. Zlobin, *Principles and methods of studying coenotic plant populations* (Kazan, 1989)
6. G.P. Semenova, *Rare and endangered flora species of Siberia: biology and protection* (Novosibirsk, 2007)
7. U.K. Laemmli, Nature, **4** (1970)
8. O.V. Agafonova (Dorogina), A.V. Agafonov, Dep. All-Russian Institute of Scientific and Technical Information, 2467-B91 (1991)
9. H. Nybom, Mol. Ecology, 13 (2004)

BIO Web of Conferences **16**, 00009 (2019) https://doi.org/10.1051/bioconf/20191600009
*Results and Prospects of Geobotanical Research in Siberia*

# Prodromus of vegetation of Yakutia

*Paraskovia* Gogoleva[1*], *Mikhail* Cherosov[2], *Svetlana* Mironova[1], *Evdokia* Burtseva[1], Nadezhda Sleptsova[1], *Boris* Pestryakov[1], *Praskovia* Kharlampyeva[1], *Lyudmila* Gavrilyeva[1], *Sargylana* Poiseeva[1], *Nikolai* Ermakov[3], *Mikhail* Telyatnikov[3], *Nikolai* Lashchinsky[3], *Elena* Troeva[2], *Evgeny* Nikolin[2], and *Viktoria* Filippova[2]

[1]M.K. Ammosov North-Eastern Federal University, 58 Belinsky Str., Yakutsk, 677000, Russia
[2]Institute for Biological Problems of Cryolithozone, Siberian Branch of the Russian Academy of Sciences, 41 Lenin Ave., Yakutsk, 677980, Russia
[3]Central Siberian Botanical Garden, Siberian Branch of the Russian Academy of Sciences, 101 Zolotodolinskaya Str., Novosibirsk,463090, Russia

**Abstract.** A brief history of development of the Prodromus vegetation of Yakutia is given. The characteristic of classes is given in a table.

The use of Braun-Blanquet approach for classification of Yakutia vegetation was invigorated by Boris Mirkin, Professor of the Bashkir State University. In Yakutia, he guided the steps of a number of specialists of the Yakut State University in their post-graduate study of botany and landscapes. Under his scientific supervision, K.E. Kononov defended his doctoral dissertation on vegetation of the Lena River floodplain [1-5], E.I. Burtseva studied the meadows on saline soils of the Lena River valley [3, 4], P.A. Gogoleva studied the vegetation of alas landscapes of the Lena-Amga Interfluve [2, 4, 5, 6, 7, 8], and S.I. Mironova also studied the alas vegetation of the Lena-Vilyuy Interfluve as well as recovery of vegetation of technogenic landscapes [2, 7, 9, 10, 11].

The students and followers of Prof. Mirkin used Braun-Blanquet approach for classification of weed vegetation (N.P. Sleptsova) [12, 13], a number of northern taiga and tundra plant communities (B.N. Pestryakov) [14-16], vegetation of the Vilyuy River basin (S.I. Poiseeva) [17-19], ruderal vegetation (M.M. Cherosov) [8, 20, 21, 22], grazing land vegetation (L.D. Gavrilyeva) [10, 23, 24], forest and mountain vegetation (N.B. Ermakov, E.G. Nikolin) [25-29]. Recently, the works on tundra vegetation classification were conducted (M.Yu. Telyatnikov, N.N. Lashchinsky, E.I. Troeva) [30-33]. Aquatic vegetation classification is underway now (P.I. Kharlampyeva, V.A. Filippova) [34-35].

The long-term work yielded the Prodromus of vegetation of Yakutia. Its structure is represented now by the following units: 36 classes, 72 orders, 78 alliances, 216 associations, 112 subassociations, 63 variants, and 13 derivative and basal communities (Table).

As it is seen from the Table, riparian, meadow and anthropogenic vegetation is most properly studied. The higher rank units are given according to Ermakov (2012) [36].

* Corresponding author: sedum@mail.ru

**Table.** Structure of the Prodromus of vegetation of Yakutia

| Class | Ord | All | Ass | Subass | Var/co |
|---|---|---|---|---|---|
| *1* | *2* | *3* | *4* | *5* | *6* |
| ***Lemnetea*** | 1 | 1 | 5 | | |
| ***Potametea pectinati*** | 2 | 3 | 12 | | |
| ***Utricularietea intermedio-minoris*** | 1 | 1 | 1 | | |
| ***Littorelletea*** | 1 | 1 | 1 | | |
| ***Phragmiti-Magno-Caricetea*** | 5 | 10 | 37 | 10 | 1 d.c. |
| ***Thero-Salicornietea*** | 1 | 1 | 2 | | |
| ***Bolboshoenetea maritimi*** | 1 | 2 | 3 | | |
| ***Scorzonero-Juncetea gerardii*** | 2 | 3 | 9 | 18 | 5 |
| ***Scheuchzerio-Caricetea nigrae*** | 3 | 5/1 | 17 | 1 | 4 |
| *?* | 1 | 1 | 1 | 2 | |
| ***Oxycocco-Sphagnetea*** | 1 | 1 | 1 | 2 | 1 |
| ***Thlaspietea rotundifolii*** | 1 | 2 | 2 | 2 | |
| ***Salicetea herbaceae*** | 1 | 1 | 4 | | 1 |
| ***Carici rupestris-Kobresietea bellardii*** | 1 | 1 | 2 | 3 | |
| ***Loiseleurio-Vaccinietea*** | 1 | 3 | 11 | 8 | 2 |
| *?* | 1 | 1 | 1 | | |
| ***Juncetea trifidi*** | 1 | 1 | 1 | | |
| ***Hylocomio-Salicetea glaucae*** | 1 | 1 | 1 | | 1 |
| ***Molinio-Arrenatheretea*** | 2 | 3 | 7 | 5 | 7/1 |
| ***Calamagrostetea langsdorffii*** | 1 | 1 | 7 | | |
| ***Cleistogenetea sguarrosae*** | 2 | 8/1 | 17 | 13 | 2/1 |
| *?* | 1 | 1 | 2 | 2 | |
| ***Vaccinio-Piceetea*** | 5 | 6 | 20 | 4 | 4 |
| ***Rhytidio-Laricetea sibiricae*** | 3 | 4 | 4 | | |
| ***Polygono arenastri-Poetea annuae*** | 1 | 1 | 1 | 2 | |
| ***Stellarietea mediae*** | 2 | 4 | 7 | 13 | /2 |
| ***Artemisietea vulgaris*** | 2 | 3 | 5 | 9 | 13/3 |
| ***Polygono-Artemisietea austriacea*** | 1 | 1 | 4 | 2 | |
| ***Epilobietea angustifolii*** | 1 | 1 | 2 | | /4 |
| ***Bidentetea tripartiti*** | 1 | 1 | 2 | 2 | |
| ***Puccinellio-Hordeetea jubati*** | 1 | 1 | 16 | 10 | 13 |
| ***Matricario-Poetea arcticae*** | 1 | 1 | 6 | 2 | 2/1 |
| ***Plantaginetea majoris*** | 2 | 2 | 3 | | |
| ***Salici-Betuletea nanae*** | 1 | 1 | | | /1 |
| ***Salicetea purpureae*** | 1 | 1 | 2 | 4 | /7 b.c. |
| **Total** **36** | **72** | **78** | **216** | **112** | **63/13** |

## References

1. K.E. Kononov, *Ecology of meadow vegetation of the Lena River floodplain*, Diss. (M, 1982)
2. K.E. Kononov, P.A. Gogoleva, S.I. Mironova, B.M. Mirkin, *Classification of vegetation of the USSR my means of floristic criteria* (M, 1986)
3. B.M. Mirkin, E.I. Burtseva, K.E. Kononov, Biol. nauki, **9** (1984)

4. B.M. Mirkin, K.E. Kononov, P.A. Gogoleva, E.I. Burtseva, L.G. Naumova, Folia Geobot. Phytotaxon, **27** (1992)
5. B.M. Mirkin, P.A. Gogoleva, K.E. Kononov, Folia Geobot. Phytotaxon, **20** (1983)
6. P.A. Gogoleva, Biol. Nauki, DEP, **246-83** (1983)
7. P.A. Gogoleva, K.E. Kononov, B.M. Mirkin, S.I. Mironova, *Sintaxonomy and simphytosociology of alas vegetation of Central Yakutia* (Irkutsk, 1987)
8. P.A. Gogoleva, M.M. Cherosov, S.G. Struchkova, E.D. Fedorova, Vestnik SVFU, **1(57)** (2017)
9. S.I. Mironova, Ecology, **5** (1983)
10. S.I. Mironova, L.D. Gavrilyeva, Z.S. Pavlova, Nauch. Dokl. Bysch.Shkoly, **2485** (1990)
11. O.I. Sumina, S.I. Mironova, Polar Geography, **27(3)** (2005)
12. N.P. Sleptsova, Bot. Zhurnal, **69(8)** (1984)
13. N.P. Sleptsova, *Weed vegetation of Central Yakutia (an experience of floristic classification application)*, Diss. (Tartu, 1985)
14. B.N. Pestryakov, P.A. Gogoleva, S.I. Mironova, DEP. AUISTI, **6289-B88** (1988)
15. B.N. Pestryakov, *Ecological-phytosociological characterization of the vegetation in the Yana River valley (northeastern Yakutia) and synanthropization processes.* Diss. (Ufa, 1991)
16. B.N. Pestryakov, P.A. Gogoleva, A.R. Ishbirdin, DEP. AUISTI, **3512-B92** (1992)
17. S.I. Poiseeva, Nauka i obrazovanie, **2** (2004)
18. S.I. Poiseeva, *Problems of the study of vegetation cover of Yakutia* (Yakutsk, Sakhapoligrafizdat, 2004)
19. S.I. Poiseeva, Fundamentalnye issledovaniya, **2** (2005)
20. P.A. Gogoleva, M.M. Cherosov, Z.S. Pavlova, B.M. Mirkin, DEP. AUISTI, **B87** (1987)
21. M.M. Cherosov, *Synanthropic vegetation of Yakutia* (Yakutsk, 2005)
22. M.M. Cherosov, N.P. Sleptsova, S.I. Mironova, P.A. Gogoleva, *Syntaxonomy of synanthropic vegetation of Yakutia* (Yakutsk, 2005)
23. L.D. Gavrilyeva, *Pasture digression and rational use of alas vegetation of the Lena-Amga Interfluve*, Diss.(Yakutsk, 1998)
24. L.D. Gavrilyeva, Sib. Ecol. Zhurnal, 18(3) (2011)
25. N.B. Ermakov, *Hemiboreal forests of continental Northern Asia (classification, ordination, coenoflora analysis)*, Diss. (Novosibirsk, 2001)
26. N.B. Ermakov, M.M. Cherosov, P.A. Gogoleva, Folia Geobot. Phytotaxon., **37** (2002)
27. N.B. Ermakov, M.M. Cherosov, Annali di Botanica, **5** (2005)
28. N.B. Ermakov, E.G. Nikolin, E.I. Troeva, M.M. Cherosov, Vestnik NGU, **7(4)** (2009)
29. N.B. Ermakov, E.G. Nikolin, E.I. Troeva, M.M. Cherosov, Vestnik NGU, **8(3)** (2010)
30. M.Yu. Telyatnikov, E.I. Troeva, P.A. Gogoleva, M.M. Cherosov, L.A. Pestryakova, S.A. Pristyazhnyuk, Flora of Asian Russia, **1(11)** (2013)

31. M.Yu. Telyatnikov, N.N. Laschinsky, E.I. Troeva, S.A. Pristyazhnyuk, P.A. Gogoleva, M.M. Cherosov, L.A. Pestryakova, Turczaninowia, **17** (2014)
32. M.Yu. Telyatnikov, E.I. Troeva, S.A. Pristyazhnyuk, P.A. Gogoleva, M.M. Cherosov, L.A. Pestryakova, Turczaninowia, **18(4)** (2015)
33. M.Yu. Telyatnikov, E.I. Troeva, S.A. Pristyazhnyuk, M.M. Cherosov, Turczaninowia, **20(3)** (2017)
34. P.I. Kharlampyeva, P.A. Gogoleva, L.A. Frolova, Scientific notes of Kazan University, **153(2)** (2011)
35. V.A. Filippova, M.M. Cherosov, E.S. Neustroeva, I.I. Osipova, Byull. MOIP (2009)
36. N.B. Ermakov, *Modern condition of basic concepts of vegetation science* (Ufa, 2012)

# Polyvariation of the development *Scutellaria* L. species and the structure of their populations in Siberia

*Alexandra* Guseva*, and *Vera* Cheryomushkina

Central Siberian Botanical Garden SB RAS, 630090 Novosibirsk, Russia

**Abstract.** Structure of coenopopulations and polyvariance of the development of species of the genus *Scutellaria* in Siberia were described. Species of the genus are semi-wood and grassy life forms. The formation of different life forms is confined to specific habitats. In a number of species (*S. galericulata, S. grandiflora* и *S. supina*), the life form is changing. Dimensional polyvariance of individuals of *S. scordiifolia* in different habitats was found. The ontogenetic structure of coenopopulations is diverse; 3 types of spectrum are formed: left-sided, centered, and right-hand. In individuals of *S. supina* and *S. tuvensis*, the duration of ontogenesis and the duration of individual ontogenetic conditions are influenced by living conditions. The polyvariance of self-maintenance was revealed in *S. supina*, the transition from completely seed self-maintenance to mixed - seed reproduction in combination with vegetative. The spatial structure and density of coenopopulations vary depending on the habitat conditions. The spatial structure of the studied species is also determined by the peculiarities of self-maintenance of coenopopulations. The change in density is also associated with the mobility of the substrate on steep, rocky mountain slopes in vegetatively stationary species and with projective cover in vegetatively mobile ones.

## Introduction

Each species in the plant community is represented by a coenopopulation. Cenopopulation has its own structure and consists of individuals differing in absolute age, ontogenetic state and other characteristics [1]. The study of various aspects of the structure of coenopopulation makes it possible to identify adaptation mechanisms that ensure their sustainable position in plant communities [2, 3].

Species of the genus *Scutellaria* L. growing in Siberia are objects of study. For this territory, 13 species are indicated [4], they grow in a wide range of habitats: from marshy and wet meadows on the plain to stony slopes and rocks in the highlands.

## Results and discussion

* Corresponding author: guseva.sc@list.ru

Resilience of coenopopulations of species of the genus *Scutellaria* in different coenoses of Siberia is ensured by population adaptations. Adaptations are manifested in a variety of life forms, changes in the dimensional characteristics of individuals, parameters of ontogenetic structure, density, type of spatial structure, pace of development and methods of self-maintenance.

*Life forms.* Species of the genus are semi-wood and grassy life forms. The genus is dominated by polycarpic terrestrial herbs [5]. Herbal biomorphs are formed mainly in species growing in swamps, meadows and steppes. In species that are common in the desert steppes and on the rocks, as a reflection of adaptation to more arid habitats, subshrub dwarf subshrub and dwarf subshrub life forms are formed. Three species in different ecological-phytocenotic conditions revealed morphological polyvariance, two or more life forms are formed [6]. Species of *S. galericulata* form 2 grassy life forms; *S. grandiflora* – herbaceous (taproot herbaceous polycarpic) and semi-woody (dwarf subshrub); in individuals of *S. supina*, 4 life forms are formed: semi-woody (dwarf subshrub and subshrub dwarf subshrub) and herbals (taproot herbaceous polycarpic and longrhizomatous-taproot herbaceous polycarpic).

*Dimensional polyvariance*. In different habitats, individuals differ in size. Thus, dimensional polyvariance was found in individuals of *S. scordiifolia* in different habitats: stepp meadow - bunchgrass stepp - meadow stepp. Analysis of average indicators of signs showed that in most of the signs, the largest are individuals growing on a stepp meadow. By most indicators, individuals of bunchgrass steppes and meadow steppes, as well as stepp meadows and meadow steppes, differ from each other. The stolone-forming life form is formed in individuals of *S. scordiifolia*, the species also belongs to vegetative annuals, it is mainly characterized by vegetative reproduction. Such morphometric indicators as the number of stolons of n + 1 order and their length give an idea of the prospects for the development of this species in specific living conditions. These parameters show its ability to consolidate and develop new territories. *S. scordiifolia*. Based on the data obtained, the most favorable living conditions for individuals of *S. scordiifolia* are steppe meadows with a high projective cover.

*Types of self-sustaining.* One of the main indicators of the state of coenopopulation is its ability to sustain self-maintenance. In most species of the dwarf subshrub life form, self-maintenance occurs only by seed. Grassy biomorphs in most cases have a mixed type of self-maintenance. In individuals of *S. supina* in different habitat conditions, a change in the type of self-maintenance occurs. Self-maintenance of coenopopulations is carried out only by seed in bunchgrass and meadow steppes. Long root shoot [7] appear in the conditions of steppe meadows, self-maintenance type mixed: seed renewal in combination with vegetative [8].

*The pace of development.* The polyvariance of the rates of individuals` development ensures the heterogeneity of individuals in coenopopulations, thereby increasing their stability in different ecological-phytocenotic conditions. The effect of habitat conditions on the duration of the ontogeny of individuals and the duration of individual ontogenetic conditions was found in *S. supina* and *S. tuvensis*. In S. supina, the temporal polyvariance is described in individuals of different life forms and in one life form under different ecological-phytocoenotic conditions. The longest life expectancy (50–55 years) was observed in the subshrub dwarf subshrub life form, the smallest life expectancy was found in taproot and longrhizomatous-taproot forms (from 21 to 30 years). In individuals of taproot life forms in different habitats, differences in the duration of the middle-aged and old generative states are observed. In *S. tuvensis*, 4 classes of polyvariance are described: normal development; slow development; accelerated development; missing one or more developmental states [9].

*Ontogenetic structure of coenopopulations.* Ontogenetic structure reflects age-related heterogeneity of coenopopulations. In general, this indicator is quite dynamic, but the study of changes in the ontogenetic structure makes it possible to identify the characteristics of the adaptation of species to the conditions of existence. Ontogenetic structure of coenopopulations of Siberian species of different life forms differs in diversity, 3 types of spectrum are formed: left-sided, centered and right-sided. In some species, the formation of several types of spectrum is noted. Thus, in *S. tuvensis*, the formation of three types of spectrum is associated with a change in the rates of development of individuals in different ecological-coenotic conditions. In individuals of *S. supina*, the formation of several types of spectrum is determined by the change of life form and the method of self-maintenance in different habitat conditions. The variety of types of spectrum, characteristic of individual species, indicates their high ability to adapt to environmental conditions.

*The spatial structure of the elementary demographic units (EDU) and the density of individuals.* EDU is many individuals of the same species, necessary and sufficient to ensure the sustainable turnover of generations in the smallest possible territory [10]. The change in the spatial structure of the ede and density in different habitats ensures the stability of coenopopulations of the species. Thus, depending on the habitat conditions, the spatial structure and density of the coenopopulations of *S. grandiflora*, *S. tuvensis*, *S. galericulata* and *S. ikonnikovii* changes. Two types of EDU spatial structure are described: discontinuous and continuous.

The continuous type of spatial structure in the vegetatively stationary species of *S. grandiflora* and *S. tuvensis* is formed on stony, gravelly slopes, pebbles and coarse scree; in narrow gorges – discontinuous type of spatial structure. The continuous type of spatial structure in the vegetatively stationary species of *S. grandiflora* and *S. tuvensis* is formed on stony, gravelly slopes, pebbles and coarse scree. The change in density (2.5–20.5 individuals per m2) is associated with the mobility of the substrate on the steep stony slopes of the mountains, and, as a result, the preservation of individual species only near large stones or on a substrate fixed by other species. A continuous type of spatial structure is formed in the meadows of vegetatively mobile species with irregular seed propagation of *S. ikonnikovii* and *S. galericulata*. On the rocky river banks – discontinuous type of spatial structure. Eremas as a result of irregular seed breeding fall into the water, which contributes to their spread along the river bank. Individual EDU of different ages are mainly concentrated around large trees. Since the species is vegetatively mobile, EDU of different ages in the population are represented by clones. The ecological density of individuals in EDU depends on the total projective cover: the higher it is, the more intense the disintegration processes and the greater the density of individuals.

Thus, the adaptation of species of the genus *Scutellaria* to ecological-phytocoenotic conditions occurs in different ways. The variety of adaptations allows the species to adapt to the constantly changing environmental conditions and to occupy a stable position in various plant communities.

The work was carried out with the financial support of the grant of RFBR within the framework of scientific project № 18-04-00621-a and project of the State Assignment of Central Siberian Botanical Garden of the Siberian Branch of the Russian Academy of Sciences № AAAA-A17-117012610053-9.

## References

1. O.V. Smirnova, L. B. Zaugol'nova and etc, *Cenopopulation of plants* (Nauka, Moscow, 1976)
2. L.B. Zaugol'nova, L. A. Zhukova and etc, *Plant cenopopulations (population biology essays)* (Nauka, Moscow, 1988)

3. L.A. Zhukova, *Population life of meadow plants* (Yoshkar-Ola, 1995).
4. V.V. Zuev, *Flora of Siberia. Pyrolaceae-Lamiaceae (Labiatae)*, **11** (Novosibirsk, 1997)
5. V.A. Cheryomushkina, A. A. Guseva, *Life forms of the species of the genus Scutellaria L. (section Lupulinaria A. Hamilt) in North and Central Asia* (Baku, 2016)
6. A.A. Guseva, V. A. Cheryomushkina, *Classification of life forms of species of the genus Scutellaria L. (Lamiaceae)* (Vladivostok, 2017)
7. L.Ye. Gatcuk, Bull. MOIP, **79(1)** (1974)
8. V.A. Cheryomushkina, A.A. Guseva, Contemporary Problems of Ecology, **8(5)** (2015)
9. A.A. Guseva, V. A. Cheryomushkina, Bull. MOIP, **122** (2017)
10. O.V. Smirnova, L.B. Zaugol'nova, R.V. Popadyuk, Biology Bulletin Reviews, **54(3)** (1993)

# The similarity and difference in the flora of the pine forest on the southern border of the range of *Pinus sylvestris* L.

*Nataliya* Kin*

Institute of Steppe of the Ural Branch of the Russian Academy of Sciences, 460000, Pionerskaya Str., 11, Orenburg, Russia

**Abstract.** The data of studies and calculations showing the high floristic similarity of the pine forests of the Russian Plain located in the southern part of *Pinus sylvestris* L. area. Especially high similarity is observed in the indigenous fraction of the flora. The similarity of floras in the adventive fraction is insignificant. Taking into account the forecast of a possible, in the near future, floristic homogenization due to the species of the adventive fraction, we calculated an estimation index, which indicates an opposite process of homogenization — differentiation. Given the division of advents into groups according to the introduction time, we calculated the estimated index separately for archaeophytes and kenophytes. It was revealed that in pairs of pine forests located relatively close to each other, archaeophytes enhance differentiation, and kenophytes enhance the homogenization of flora. Differential taxa (families, genera, and species) are found only in one of all the studied pine forests, that determine the characteristics of the studied floras. Endemic species have been established that give originality to the flora of the studied pine forests.

## 1. Introduction

Pine forests, on the southern limit of *Pinus sylvestris*, are widespread in the steppe and forest-steppe zones. The edge areal location of the pine forests makes them the most dynamic and sensitive to environmental changes, which is primarily reflected in the composition and development of the flora. Florogenesis historically had an aborigen basis [2], but at present, alien elements — adventiv species — are exerting a growing influence on this process. According to forecasts made regarding adventiv species of the flora of the Central Black Earth Region of Russia, [3] percentage of this fraction will steadily increase. This is facilitated by climate change, increased anthropogenic transformation of landscapes and, as a result, the formation of derived landscapes with altered zonal communities. The amount of adventiv species in flora increases phyto-diversity [4, 5], but over time, the resettlement of some species may lead to increase similarity between floras. This phenomenon is called "biotic homogenization" [4, 6], "unification of flora" [3, 7]. Adventiv

* Corresponding author: kin_no@mail.ru

species, increasing their ranges, can negatively affect the ranges of indigenous, mostly rare and endemic species, in the direction of their reduction [8, 9, 10].

On the contrary, the same climatic features, characterized by sharp fluctuations in the conditions of a subarid climate, the difference in the historical and geological processes of the formation of territories, the features of anthropogenic impact can limit some and catalyze the penetration of other adventiv species. This process, the opposite of homogenization, is called differentiation.

## 2. Materials and methods

In order to identify floristic similarities and differences, to establish the peculiarities of florogenesis due to the adventive fraction of the flora, we studied the pine forest flora of the Russian Plain: Usmansky (an area of 70700 hectares, 31053 hectares natural reserve) and Khrenovsky (40210 hectares) located in the forest-steppe subzone of the Oka-Don lowland [11], Buzuluksky (106788 hectares, 14399 hectares - the protection zone) and Krasnosamarsky (13554 hectares), located in the steppe zone of the Volga region [12].

The Jaccard coefficient was used to identify the similarity of pine forest flora ($K_J$):\

$$K_J = \frac{C}{A+B-C} \quad (1),$$

C is the number of species for a pair of compared pine forests; A – the number of species in the flora of one comparable pine forest; B – the number of species in the flora of another comparable pine forest.

If the coefficient $K_J$ exceeds 0.5, then more than half of the species in the flora of the studied pine forests are common and indicates a high similarity of the compared flora.

The Jaccard coefficient was calculated both for the pine forests flora as a whole and separately for indigenous ($K_{Jind}$) and adventive ($K_{Jadv}$) species, dividing the latter into archaeophytes ($K_{Jarch}$) and kenophytes ($K_{Jken}$).

To identify the role of species of the adventive fraction in the process of homogenization or differentiation at this stage of the development of pine forests flora, an estimated index is calculated. The estimated index (H) is expressed through the difference the Jaccard coefficients calculated for the floras of the studied pine forests with the participation the adventive species ($K_J$) and the indigenous component of the flora ($K_{Jind}$):

$$H = K_J - K_{Jind} \quad (2)$$

A positive result indicates homogenization, and a negative indicates differentiation of the floras [4].

Considering that advents have a temporary gradation (archaeophytes and kenophytes), it is worthwhile to find out due at the expense a group this or that process dominates. For this purpose, the estimated index for archaeophytes ($H_{arch}$) and kenophytes ($H_{ken}$) was calculated:

$$H_{arch} = K_{Jadv} - K_{Jarch} \quad (3)$$

$$H_{ken} = K_{Jadv} - K_{Jken} \quad (4),$$

a positive result indicates homogenization, and a negative one indicates differentiation of flora.

## 3. Results

The results of calculating the similarity of floras by the Jaccard coefficient, their indigenous and adventive of fractions are given in the table 1.

**Table 1.** The similarity of the flora of pine forests in the Jaccard coefficient

| Pine forests | Usmansky | | Khrenovsky | | Buzuluksky | | Krasnosamarsky |
|---|---|---|---|---|---|---|---|
| | $K_{Jadv.}$ | | | | | | |
| Usmansky | - | | ***0,4*** | | ***0,28*** | | ***0,24*** |
| Khrenovsky | 0,60 | 0,65 | - | | ***0,32*** | | ***0,33*** |
| Buzuluksky | 0,52 | 0,54 | 0,55 | 0,55 | - | | ***0,49*** |
| Krasnosamarsky | 0,43 | 0,45 | 0,49 | 0,47 | 0,52 | 0,53 | - |
| | $K_J$ | $K_{Jind.}$ | $K_J$ | $K_{Jind.}$ | $K_J$ | $K_{Jind.}$ | |

The greatest similarity of flora was found in the Usmansky and Khrenovsky pine forests located in the same natural subzone and geographically close to each other 95 km The similarity in the indigenous fractions of the flora of these forests is especially high - 0.65. A sufficiently high floristic affinity was revealed near Buzuluksky pine forest with all the other pine forests under consideration, both in general for the flora and its indigenous fraction. The high similarity of the Buzuluksky pine forest flora with other pine forests is probably associated with a significant variety of landscapes, ecosystems, unique habitats, which within this object provides a high floristic diversity of the territory, and, as a result, increases the similarity with other pine forests. In addition, the Buzuluksky and Krasnosamarsky pine forests are connected by a narrow strip of forest along the right bank of the Samara river [13] and are located in the steppe zone of the Volga region.

In the adventiv fraction of the flora of the studied pine forests, the similarity is small. The closest adventiv fractions of the flora of the Buzuluksky and Krasnosamarsky, as well as the Usmansky and Khrenovsky pine forests, which is explained by the proximity of the location of these pairs of forest areas.

The estimated index (table 2) for each pair of compared pine forests has a negative value, which indicates the process of differentiation, with the exception of the pair - Buzuluksky - Khrenovsky forests.

**Table 2.** Estimated indices of the flora of the studied burs and their adventiv fractions

| **The compared pine forests** | **The distance between the pine forests, km** | **H** | $H_{arch}$ | $H_{ken}$ |
|---|---|---|---|---|
| Krasnosamarsky - Buzuluksky | 75 | -0.01 | -0.174 | 0.109 |
| Khrenovsky – Usmansky | 95 | -0.05 | -0.176 | 0.042 |
| Krasnosamarsky – Khrenovsky | 768 | -0.02 | 0.045 | 0.095 |
| Krasnosamarsky – Usmansky | 789 | -0.02 | -0.389 | -0.571 |
| Buzuluksky – Khrenovsky | 840 | 0.00 | -0.123 | 0.095 |
| Buzuluksky – Usmansky | 865 | -0.02 | 0.053 | 0.065 |

The data obtained are consistent with the conclusions made by O.V. Morozova [4] that the greater the distance between the flora, the less differentiation. So, the relatively closely located forests Khrenovsky and Usmansky have the highest differentiation among all the studied forests. The remaining pairs of forests are at a considerable distance from each other. In this case, the homogenization coefficient increases, and the differentiation decreases. The exception is a pair of Buzuluksky - Krasnosamarsky pine forests. They are closely located (75 km), but are characterized by less differentiation of flora than the rest, located at a greater distance from each other. Perhaps this is due to the fact that the Krasnosamarsky and Buzuluksky forests in the past constituted a single massif.

Calculations showed that in pairs of pine forests located relatively close to each other due to archaeophytes differentiation occurs, and due to kenophytes - homogenization of the flora (table 2). In the Krasnosamarsky and Khrenovsky pine forests, as well as in the Buzuluksky and Usmansky, both temporary groups of adventivs are involved in the process of homogenization. This is probably due to the type of use of these forests. Khrenovsky and Krasnosamarsky pine forests do not have a high conservation status and are actively used by humans, and the Usmansky and Buzuluksky pine forests are partially or fully federal protected areas. In the Krasnosamarsky-Usmansky pine forests pair, the process of differentiation proceeds both due to archaeophytes and kenophytes.

When comparing the flora of pine forests, differential species, genera, and families were revealed — taxa that reflect the nature of the flora and are found only in one of the compared floras [14]. The data obtained are summarized in table 3. Undoubtedly, the presence of such taxa most clearly emphasizes the peculiarity of the studied floras.

**Table 3.** Quantitative indicators of the differential species of the studied pine forests

| The name of pine forests | 1 | 2 | 3 | 4 |
|---|---|---|---|---|
| Absolute number of differencial species / share (%) of all recorded species in the flora | 249 / 23.2 | 47 / 6.0 | 94 / 12.0 | 56 / 8.5 |
| Aborigens, their absolute number / share (%) of all indigenous species recorded in the flora | 148 / 17.4 | 39 / 5.8 | 72 / 10.7 | 51 / 8.8 |
| Adventivs, their absolute number / share (%) of all adventivs species in the flora | 89 / 41.8 | 8 / 7.1 | 18 / 16.5 | 4 / 5.1 |
| - absolute number of archaeophytes / share (%) of all archaeophytes recorded in the flora | 11 / 26.8 | 2 / 6.7 | 3 /8.8 | 0 / 0.0 |
| - absolute number of kenophytes / share (%) of all kenophytes recorded in the flora | 78 / 45.3 | 6 / 7.3 | 15 / 20.0 | 4 / 8.3 |
| Introducers, their absolute amount / share (%) of all introducers in the flora | 12 / 100.0 | 0 / 0.0 | 4 / 57.1 | 1 / 33.3 |

**Pine forest**: 1. Usmansky, 2. Khrenovsky, 3. Buzuluksky, 4. Krasnosamarsky.

In the Usmansky pine forest has the largest number of differential species, but only 59.4% of the aborigines among them. In other studied pine forests the share of indigenous differential species is more significant and makes 76.6 % in Buzuluksky, 82.9 % in Khrenovsky and 91.0 % in Krasnosamarsky pine forests. Correspondingly, the share of differential adventivs species in the Usmansky forest is more significant (35.7 %), compared with other studied forests, where their percentage of 7.1 is in Krasnosamarsky, 17.0 in Khrenovsky and 19.0 in Buzuluksky. It should be noted that in all pine forests under study, among adventiv differential species, the highest percentage of kenophytes, with a later skidding, compared with archaeophytes.

The flora of the Usmansky pine forest has the largest number of differential taxa. 12 families are noted only in the flora of this forest. Of these, 4 families include exclusively aborigen species: *Ericaceae, Oxalidaceae, Elatinaceae, Saxifragaceae,* the remaining 8 contain representatives of the adventiv fraction of the flora. In the flora of the Usmansky pine forest, 72 differential genera were registered, among which there were 37 genera from the aborigens, 34 from the adventiv fraction of the flora and 1 genus from introducent. 5 endemic species recorded in the pine forest flora: *Euphorbia gmelinii* Steud., *Psephellus sumensis* (Kalen.) Greuter, *Stipa anomala* P. Smirn s.str., *Aconitum nemorosum* Bieb. ex Reichenb.

In the flora of Khrenovsky pine forest, the *Adoxaceae* family is differential. Like *Adoxa*, the genera *Aconogonum, Ononis, Berula* are differential for this pine forest. Unlike Usmansky, in the flora of Khrenovsky forest no differential families and genera from the adventitious fraction of the flora were revealed. Originality of the flora of Khrenovsky pine forest is given by 4 endemic species: *Centaurea pineticola* Iljin, *Centaurea sophiae* Klokov., *Iris pineticola* Klokov., *Festuca wolgensis* P.A. Smirn.

In the flora of the Buzuluky pine forest, one differential family of *Ephedraceae* and 8 genera was revealed. Among the latter, 5 contain representatives of the aborigen fraction of pine forest flora. 8 endemic species identified here: *Lathyrus litvinovii* Iljin, *Elytrigia reflexiaristata* (Nevski) Nevski, *Hylotelephium zhiguliense* Tzvelev, *Astragalus wolgensis* Bunge, *Tragopogon tanaiticus* Artemcz., *Dianthus fischeri* Spreng., *Myosotis popovii* Dobrocz., *Chondrilla graminea* M. Bieb.

The flora of Krasnosamarsky pine forest differs from others in the presence of the *Limoniaceae* family and 9 genera that are differential for it. Of these genera, 7 combine the species of the aborigen fraction of the flora. The special nature of the flora of Krasnosamarsky pine forest was expressed in the presence of 2 endemic species: *Ferula caspica* Bieb. and *Allium podolicum* (Asch. & Graebn.) Blocki ex Racib.

## 4. Discussion

The result of the research showed high floristic similarity between the forests of the Russian plain. As flora is represented by two factions, noted that the similarity in of the aborigen faction is much higher than in the adventive fraction of flora. The predictions of scientists about the approaching unification of the flora due to the active adaptation and settlement of alien species was not confirmed to us at this time. This is also indicated by the data obtained in the calculation of the estimated indices of the compared pine forests It was established that kenophytes have a greater effect on homogenization, and archaeophytes on the differentiation of flora. The manifestation of this or that process significantly depends on the presence or absence of the security status of the studied pine forests.

The most striking floristic differences of the studied pine forests were shown due to differential taxa and endemic species.

## 5. Acknowledgements

The work was carried out in the framework of the state task of Institute of Steppe of the Ural Branch of the RAS, theme № AAAA-A17-117012610022-5

## References

1. E.A. Starodubtseva, Vascular plants. *Flora of the Voronezh reserve. Vascular plants, bryophytes, lichens, mushrooms. Flora and fauna of reserves.* (Moscow, 1999)
2. A.I. Tolmachev, *Comparative floristry methods and problems of florogenesis* (Novosibirsk, 1986)
3. A.A. Tishkov, *Adventive flora of the Voronezh region: Historical, biographic, economic aspects* (Voronezh, 2004)
4. O.V. Morozova, Russian Journal of Biological Invasions, 3 (2018)
5. D.F. Sax, S.D. Gaines, Trends in Ecology and Evolution, **18**, 11 (2003)
6. M.L. McKinney, J.L. Lockwood, Trends in Ecology and Evolution, **14** (1999)
7. A.A. Tishkov, Anthropocene, Proceedings of the Zoological Museum of Moscow State University, **54** (2016)
8. J.D. Olden, N.L. Poff, Am. Nat., **162 (4)** (2003)
9. H. Qian, R.E. Ricklefs, Ecology Letters, **9** (2006)

10. M. Winter, I. Kühn, F.A. La Sorte, O. Schweiger, W. Nentwig, S. Klotz, Global Ecology and Biogeography (2010)
11. F. N .Milkov, Natural areas of the USSR (Moscow, 1977)
12. Buzuluk Bor. *General essay and forest crops*, **I** (Moscow-Leningrad, 1949)
13. E.S. Korchikov, N.V. Prokhorova, T.I. Plaksina, N.M. Matveev, Yu.V. Makarova, A.I. Kozlov, Samara Luka: problems of regional and global ecology. Togliatti, **19**, 1 (2010)
14. R.V. Kamelin, *Flora of the North of European Russia (in comparison with the neighboring territories)* (Saint-Petersbourg, 2017)

BIO Web of Conferences **16**, 00012 (2019) https://doi.org/10.1051/bioconf/20191600012
*Results and Prospects of Geobotanical Research in Siberia*

# The status of some herbaceous species populations in West Siberian subtaiga forests

*Elizaveta* Komarevtseva*, *Natalia* Kurochkina, and *Natalia* Gordeeva

Central Siberian Botanical Garden of the SB RAS, 630090 Novosibirsk, Russia

**Abstract.** The coenopopulations of *Bupleurum longifolium* L., *Pulmonaria mollis* Wulf. ex Hornem., *Angelica sylvestris* L were studied in the territory of the experimental forest park of CSBG (Novosibirsk). The coenopopulations have a left-sided spectrum. The main method of reproduction of species – seed. In the intact pine-birch forest, coenopopulations of species with a low density of individuals develop. In weakly disturbed habitats (selective felling, periodic haying) the densities of individuals in *B. longifolium*, *P. mollis* increases, as well as old (in *B. longifolium*) or juvenile (in *A. sylvestris*) individuals fall out. Improving lighting here causes an increase in the size parameters of generative individuals and the acceleration of their development.

To estimate vegetation wellbeing of specially protected natural object «Central Siberian Botanical Garden (CSBG SB RAS)», coenopopulations of *Bupleurum longifolium* L., *Pulmonaria mollis* Wulf. ex Hornem., *Angelica sylvestris* L were investigated in July-August 2018. These species are widespread in small-leaved and light-coniferous subtaiga forests of West Siberia [1; 2]. To reveal the status of coenopopulations these species, their ontogenetic structure and peculiarities of the development of generative plants in various growing conditions were studied.

The ontogenesis was described: *B. longifolium* [3; 4], *P. mollis* [5], *A. sylvestris* [6]. Structure of coenopopulations was studied according to the common technique [7; 8]: 10–42 plots of 1x1 m$^2$ were set; specimens of each ontogenetic state were counted. The number of plots depended on the abundance of species in the cenosis. 10–25 generative individuals were examined at coenopopulations to identify developmental features.

Forest Park of the CSBG is situated on right bank of Ob' river. Pine-birch forests are the background forest type; most of territory is intact, but some sites are subjected to selective tree felling or are crossed by country roads. We choose three sites. **The first site** (site 1) is located in intact pine-birch herb forest: cover of *Betula pendula* Roth is 60 %, *Pinus sylvestris* L. is 5 %. Total cover of herbs is 80 %: *Pteridium aquilinum* (L.) Kuhn (40%) and *Aegopodium podagraria* L. (10 %) are dominate. Grass cover is made by *Brachypodium pinnatum* (L.) Beauv. and *Carex macroura* Meinsh. In this site, a study was conducted on all species with a low coverage here: *B. longifolium*–5 %, *A. sylvestri* –1 %, *P. mollis*–1 %. The other two sites are exposed to anthropogenic influence. **The second site**

* Corresponding author: elizavetakomarevceva@yandex.ru

(site 2) situated on the edge of grass birch forest, subjected to selective felling: cover of *Betula pendula* is 50%. Grasses prevail: *Brachypodium pinnatum* (12 %), *Dactylis glomerata* L. (7 %), *Calamagrosti arundinacea* (L.) Roth (7 %), *Carex macroura* (12 %). Herbs have a little less cover: *Aegopodium podagraria*–12%, *Heracleum dissectum* Ledeb.–10 %, *P. mollis*–10 %, *Veronica chamaedrys L.*–5 %. With the exception of *P. mollis*, the cover of the studied species is low (1–2 %). On this site we studied *P. mollis* and *B. longifolium*. **The third site** (site 3): forest meadow with shrubs is located next to country road, the edges of which are periodically mowed. The shrub layer is composed of *Salix cinerea* L. (25 %) and juvenile *Tilia cordata* Mill. (15 %). Total herb cover is 80 %, cover of grasses is low (10 %): *Dactylis glomerata* (5 %), *Carex macroura* (3 %), *Poa angustifolia* L. (3 %). *Veronica chamaedrys* (30 %), *Aegopodium podagraria* (20 %), *A. sylvestris* (20 %), *Filipendula ulmaria* (L.) Maxim (10 %) dominate in herb layer. *Cirsium helenioides* (L.) Hill (5 %) and *C. setosum* (Wild.) Bess. (2 %) indicate anthropogenic influence. On this site we studied *A. sylvestris*; *B. longifolium* and *P. mollis* were not abundant (1–3).

*B. longifolium*, *P. mollis* belong to semirosette short-root polycarpics. Individuals of species have a complex ontogenesis, in which the seed individual (genet) is divided into several daughters (ramets). Fragmentation (disintegration) of individuals occurs in the generative state. It is accompanied by a shallow rejuvenation of ramet in *B. longifolium* [3]. In *P. mollis* ramet rejuvenation does not occur, the formed particulates (ramets) quickly age and pass into the old generative state [5]. Both species form an implicit polycentric life form with non-specialized complete disintegration [9; 10]. *A. sylvestris* is a perennial monocarpic plant. During simple ontogenesis (vegetative reproduction is absent) the species develops a monocentric biomorph. Regardless of the ability to vegetative reproduction, resumption of coenopopulations in all three species occurs by seed. So their coenopopulations here have left-sided spectrum (Fig. 1).

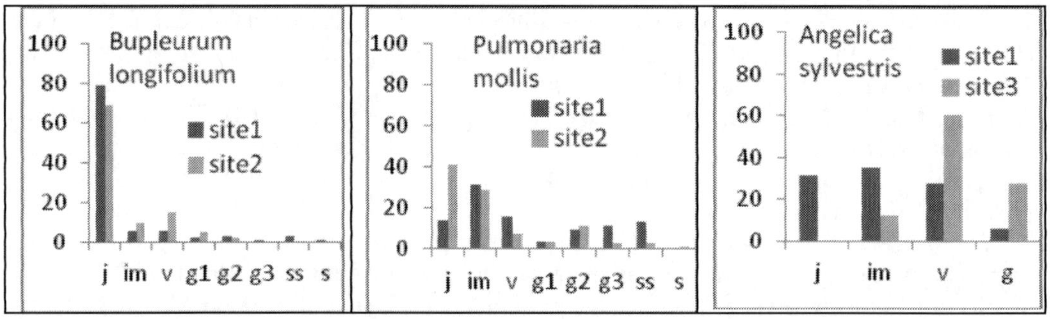


**Fig. 1.** Ontogenetic spectra of species. On the X-axis– ontogenetic states; on the Y-axis – the percentage of individuals.

In the spectra of *B. longifolium* is dominated by juvenile individuals (68.8–79.4 %). Immature (*im*) and adult vegetative (*v*) plants are represented much less: (5.6–9.5 %). The presence of all generative individuals is also insignificant: from 6.1 % to 13 %. All this testifies the effective seed renewal of this species in forests regardless of the degree of habitat disturbance. A selective felling in the birch forest (site 2) causes the loss of old generative ($g_3$) and post-generative individuals (*ss*, *s)*, as well as an increase in seedlings (up to 24 pieces/m$^2$). As a result, the density of coenopopulation increases to 21.3 PCs /m$^2$ compared to the 1st site (16.8 PCs /m$^2$).

Habitat conditions significantly affect the development of generative plants. Adult individuals *B. longifolium* form shoots of different cyclicity: monocyclic (annual) and polycyclic (2–5-year). In the dark coniferous forests of Salair monocyclic shoots develop in middle-aged generative individuals ($g_2$), polycyclic shoots – mainly in young ($g_1$) and old

($g_3$) generative [4]. In the forest Park of CSBG monocyclic generative shoots develop in young and mature generative ($g_2$) states. At the end of the generative period, mainly di - and tricyclic (2–3-year) shoots are formed. Monocyclic shoots develop only from the renewal buds on the basal part of the shoot, and di- and tricyclic shoots – often from dormant buds on the rhizome. There is an increase in the number of monocyclic shoots in individuals at the edge of the birch forest (52 % of the total number of generative shoots) compared with the pine-birch forest (39 %). Comparison of morphological parameters of generative individuals from coenopopulations showed that at the edge (site 2), in better light individuals *B. longifolium* are large in size: the length of the shoot is an average of 124.6 cm, its weight – 4.8 g. In undisturbed forest (site 1) these parameters below: 119 cm and 3.4 g. This allows concluding that an improvement in the light regime has a positive effect on the development of *B. longifolium*. At the population level, this is expressed in an increase in the number of seedlings and the density of coenopopulation, and at the organizational level – in reducing of the generative shoot cyclicity and increasing its size.

Coenopopopulations of *P. mollis* have a left-sided spectrum: the main maximum is formed on juvenile or immature plants, the 2nd maximum – on adult or old generative. In the birch forest (site 2) the maximum on juvenile individuals (41.2 %) is due to the active settling of the species at partial felling of forest areas: up to 13 seedlings per 1m$^2$. The majority of young vegetative individuals (*j, im*) are eliminated at the transition to the next age state (*v* – 7.3–16.1%). The increase in seedlings and juvenile individuals (j) under disturbed grass causes a significant density of coenopopulations of *P. mollis* in birch forest (site 2) to 20.6 PCs/m$^2$ compared with pine-birch (site 1: 3.4 PCs/m$^2$).

Morphological parameters of mature generative individuals of *P. mollis* change in different communities. At the edge of the birch forest (site 2) flowering plant consists of 1–5 rosette and 1–6 generative shoots length 19.5–56.5 cm. In pine-birch forest (site 1) generative plant has 1–4 rosette and 1–4 generative shoots reaching a length of 11.5–49.5 cm. This leads to a noticeable difference in the above-ground mass of one plant: in the birch forest up to 14.7 g and in the pine-birch – up to 6.6 g. But most significantly differ in plants by mass of rhizomes. In the birch forest in disturbed cutting areas is the development of powerful rhizomes in mature generative individuals: diameter up to 1.5 cm and weight up to 24 g. Rhizomes are thinner (up to 0.8 cm), their weight reaches up to 4.2 g. in the pine-birch forest. The development of the generative part of the plant is the same in both habitats. The formation of powerful rhizomes allows the species to keep its territory in disturbed habitats. In addition, there is a slight acceleration in the development of the apical bud of the rosette shoot. So in the pine-birch forest in the apical bud there are 3–4 leafy rudiments in mid-June, in a birch forest up to 6 rudiments and in a few cases even the rudiments of the generative. It should be noted that in the pine-birch forest, many apical buds are damaged by larvae of the May beetle and the next year the elongated shoots of them will not develop, and this will cause a break in the flowering of many generative individuals. So, coenopopopulations of *P. mollis* have a left-sided spectrum. In the pine-birch forest, less developed individuals form; possible breaks in flowering due to damage to the apical buds rosette shoot larvae of the May beetle. On the edge of the birch forest increases the efficiency of seed renewal species, the rate of development of individuals slightly accelerated, as well as forming powerful rhizomes in generative plants, which allows the species to keep the territory.

*A. sylvestris* develops during simple ontogenesis, which ends in a generative state. The left-sided spectra of *A. sylvestris* are characterized by the absence of old non-flowering (*s, ss*) plants due to its monocarpicity. Ontogenetic spectra of coenopopulations differ in the position of the maximum, the proportion of generative plants, and the degree of completeness. So in the pine-birch forest cenopopopulations are full-range (all ontogenetic groups are included), it is dominated by immature individuals (35.2 %) with a small number of generative (5.9 %). On the bushy forest meadow, near the country road (site 3), the peak falls on adult vegetative plants (60.3 %), and significantly more generative individuals (27. 6 %) in the complete absence of juvenile. Probably, due to the mowing, the generative individuals do not have time to get disseminated, and as a result a juvenile group falls out. About haymaking evidence we found stumps of cut last year's generative shoots *A. sylvestris*. Differences in the spectrum do not affect the density of the species (3–3.6 PCs/m2). The greatest development is reached by generative individuals in the bushy forest meadow (site 3). The average values of the basic morphological parameters of flowering plants are higher here, which is explained by the best lighting with sufficient soil moisture.

Thus, in the Forest Park of the CSBG the coenopopulations of *Bupleurum longifolium*, *Pulmonaria mollis* and *Angelica sylvestris* are in a satisfactory condition in the intact and slightly disturbed habitats. Coenopopulations of the species have a left-sided spectrum, which is due to the predominance of seed renewal in them. Normal full-range cenopopulations with a low density of individuals develop in the bracken-grass pine-birch forest (an intact site). Growth in habitats with a small anthropogenic impact (selective felling, periodic mowing) increases the efficiency of seed renewal in *B. longifolium*, *P. mollis*, which causes an increase in the density of cenopopulations of these species due to pregenerative plants (*j*, *im*). But at the same time, it is possible the loss of old (*B. longifolium*) or juvenile (*A. sylvestris*) plants from the coenopopulations. Habitat conditions significantly affect the development of generative individuals of these species. Lighting improvement while maintaining sufficient soil moisture in slightly disturbed habitats contributes to the increase in the size characteristics of adults in all three species. In *P. mollis* and *A. sylvestris*, an increase in the biomass of above-ground part of plants is also observed by 1.5–2 times, which causes and increase in the coverage of these species to 10–20 % in slightly disturbed cenoses. In addition, under these conditions, there is acceleration in the rate of development of adult individuals. In *B. longifolium*, this is expressed in an increase in the number of monocyclic shoots in generative individuals, in *P. mollis* – in the formation of a larger number of rudiments in the apical bud of the rosette shoot.

## References

1. M. Pimenov, *Flora of Siberia*, **10** (1996)
2. O. Nikiforova, *Flora of Siberia*, **11** (1997)
3. L. Vasilyeva, N. Lashchinsky, Plant resources, **3** (1987)
4. E. Podgaevskaya, Ontogenetic atlas of plants, **3** (2002)
5. T. Petrova, Ontogenetic atlas of plants, **3** (2002)
6. E. Vinokurova, *Terpenoids in ontogenesis Angelica sylvestris L. in dark coniferous forests of Salair* (Barnaul, 2013)
7. A.A. Uranov, Biol. Nauki, 2 (1975)
8. O. Smirnova *Cenopopulations of plants (principal concepts and structure)* (M, Nauka, 1976)
9. L. Gatzuk., O. Smirnova , L. Zaugolnova, L. Zhukova, J. Ecol., **68** (1980)
10. O. Smirnova, M. Palenova, A. Komarov, Russian J. of Developmental Biology, **33** (2002)

# On the flora of honey plants of the Northern and Northwestern Altai regions within the Altai Territory

*Tatyana* Kopytina*, *Galina* Nenasheva, and *Marya* Ivanova

Altai State University, Lenina pr., 61, Barnaul, 656049, Russian Federation

**Abstract.** The revision of honey plants in the regions' floras and nature objects is regularly made for estimation of melliferous capacity of the territory. The article is devoted to the investigation of the melliferous flora of Northern and North-Western Altai in the area of Altai Territory. The goal of the research was to reveal the botanical composition of authentic honeys. The representativeness of melliferous species among common floristic richness in the studied districts is the following: in Altaiskiy – 56.49 %, Smolenskiy – 60.17 %, Soloneshenskiy – 53.61 %, Charyshskiy – 44.57 %. The examples of the botanical content of samples typical for each of these districts have been shown. In the angelica (djagilevyj) honey from Altajskiy district the maximum content of pollen belongs to the representatives from Apiaceae family 78.86 %. For the angelica (djagilevyj) honeys from surveyed districts the high content of pollen from the families Asteracerae, Brassicaceae, Fabaceae, Rosaceae are typical with different percentage in the samples. The article provides information on examples of wild-growing plant species including relicts, potential melliferous which pollen can be used as "marker" for determination of honey as Altai-origin.

Honey plants were actively examined in 20 century; nowadays the interest to this theme is still alive. The importance of the melliferous group studying is connected with the wish of customers to know the botanical composition of honey and its origin with the aim to identify falsifications.

The information on the species content of this group is regularly enriched thanks to publications. For example, A. N. Burmistrov, V. A. Nikitina [1] gave the characteristic of 88 melliferous plants. M. K. Hushnazarov [2] reported that in vegetation cover of the reserve "Romit" (Tajikistan) from 1240 wild plant species the fifth part are milleferous or beebread flowers. In the reserve "Shulgan-tash" (Bashkiria) melissopalynological analyses of honey, pollen pellet, beebread have been made. As a result, 187 milleferous and beebread plant species were revealed [3].

For the purpose of investigation of botanical and geographical origin of Russian honeys R. G. Kuminov, A. R. Ishbirdin [4] conducted the palynological analyses of 119 samples of honey from 31 regions of Russia. The content of studied samples included pollen of 183

*Corresponding author: tatkop70@mail.ru

taxa from 51 families. The authors presented photos of 40 pollen of honey species identified in samples.

In the rich and divers flora of the Russian Far East 90 primary, 100 secondary and more than 100 weak melliferous plants have been revealed. The most valuable are all species of *Tilia*, *Acer*, *Salix*, *Phellodendron amurense* Rupr., *Padus maackii* (Rupr.) Kom., *Kalopanax septemlobus* (Thunb.) Koidz., *Aralia elata* (Miq.) Seem., *Eleutherococcus senticosus* (Rupr. et Maxim.) Maxim., *Lespedeza bicolor* Turcz., *Rubus crataegifolius* Bunge etc. [5].

Melliferous capacity of the territory is mostly depended on the area on which the honey plants grow and less on the diversity of species content. V. N. Kulakov [6] examined melliferous sources in Russian federation (RF) and payed attention that there are melliferous lands the contribution of which in the honey potential is high. For example, the vegetation of burnt areas, felling sites, open forests etc. produces 2.5 milliards kg of honey (22 % of the whole deposit of honey in the country). These lands occupy a vast territories (101.7 mil. ha), but have less melliferous capacity. Moreover, big part of those lands is situated in the territories that hard to reach with not always suitable climatic conditions for bee farming. The plantings of entomophile cultures occupy 27.6 thousand ha (4 % from all of melliferous cultures of RF) and produce 1.7 milliards kg, that is 15 % from all potential deposit of RF. The highest value in creation of the melliferous storage belongs to the forest and other natural melliferous vegetation sources – 84 % from the total maximum potential. Altai Territory has high relative input of agricultural crops in potential honey deposit – 88 % [6].

Determination of geographical origin of honey is based on that fact that the whole spectrum of pollen corresponds with the flora of the region in which the honey is gathered and with the standard spectrum or descriptions of botanical content, published in the literature [7]. For the identification of the honey from Spain, for example, can be used the results of the following investigation. O. Escuredo, M. Fernández-González, M. C. Seijo [8] carried out melissopalynological analyses of 86 honey samples, gathered in north-western part of Spain, and revealed 90 types of pollen. *Castanea sativa*, *Cytisus*-type and *Rubus* occurred in 100 % of samples, as well as *Erica* in 97,7 % and *Eucalyptus* in 86 %. In addition, the pollen of *Echium*, *Quercus* и *Trifolium*-type frequently appeared. Pollen of *Salix*, *Crataegus monogyna*-type, *Campanula*-type, *Conium maculatum*-type, *Frangula alnus*, *Lotus*-type, *Lithodora* and *Myosotis* comprised 3 % of the pollen spectrum and had a big value in honey. Fifty types of pollen occurred in several samples (less than 10 %).

The present study was aimed to examine the melliferous flora of Northern and North-Western Altai in the area of Altai Territory on the territories of four administrative districts. It has a goal to reveal plant species which pollen existence characterize authenticity of local honeys. According to the scheme of botanic-geographical zonation of Altai Territory, the territory of Altaijskiy, Smolenskiy, Soloneshenskiy districts belongs to *North Altai*, and Charyshskiy – to the *Northwesternaltai* botanic-geographical regions in *North-Altai-North-Westernaltai* floristic area. The vegetation cover of those districts is characterized by diversity of vegetation types from low- to uplands: steppes, meadows, mires, fir and mixed forests, riparian and littoral vegetation.

V.I. Vereschagin [10] have written that there are different formation of forests, meadows and wetlands in Altai Territory, which are different in milleferous plants' composition. The list of milleferous (nectariferous, polliniferous) plants is compiled based on the published data [4, 5, 9–11]. We give status of milleferous to the species which nature was confirmed by the bee farming experience, and, in addition, by examine of the flower's structure and existence of pollen from mellitophilae plants in honey. For the identification of pollen in the samples of honey, we compare it with the pollen from flowering plants and from taken herbarium specimen. There are some important conditions for the good quality identification of pollen in honey. They are resolving power of the microscope and existence of preserved pollen samples of identified milleferous species. In 2015, we created the collection of recent pollen of 341 honey plant species and 103 specimen of honey. For the pollen identification we used works of L. V. Kupriyanova, L. A. Aljoshina [12], R. G. Kurmanov, A. R. Ishbirdin [4], Austrian database of the project PONET (Pollen, https://www.ages.at/themen/umwelt/pollen/projekt-ponet/) [13], containing the high resolution images of pollen of 800 plant species from regional and adventive flora.

According our data the group of melliferous plants in examined districts includes in Altaiskiy – 326 species, Smolenskiy – 343, Soloneshenskiy – 364, Charyshskiy – 415 [14]. However, representativeness of melliferous species on the background of floristic diversity in districts is another: Altaiskiy – 56.49 %, Smolenskiy – 60.17 %, Soloneshenskiy – 53.61 %, Charyshskiy – 44.57 %. Such occurrences can be explained by that fact that in Soloneshenskiy and Charyshskiy districts vegetation communities are more different with many species that not presented in two other districts, especially species that are not melliferous. The melliferous qualities are not examined for many species of high lands.

We have made the calculation of frequency of the botanical taxa pollen among 1000 pollen grains in each specimen made from honeys from each of four districts. We will show the examples of the most representative samples of honey. In the specimen of angelica (djagilevyj) honey from Charyshskiy district the high content of pollen from the families Apiaceae (59.92 %), Brassicaceae (11.98 %), Fabaceae (8.26 %) и Rosaceae (7.29 %) is presented. In addition, in small concentration the pollen grains of the following families have been found Valerianaceae (5.83 %), Asteraceae (4.86 %), Polygonaceae (1.62 %), Pinaceae (0.16 % – occasional import of pollen) and Tiliaceae (0.08 % – *Tilia* is growing frequently in artificial plantings around apiary which the sample of honey was taken from).

In the sample of angelica (djagilevyj) honey from Soloneshenskiy district the pollen from Apiaceae (76.54 %), Brassicaceae (10.26 %), Asteracerae (3.62 %), Fabaceae (3.36 %), Lamiaceae (2.86 %), Rosaceae (2.44 %), Polemoniaceae (0.50 %), Liliaceae (0.34 %) and Polygonaceae (0.08 %) families has been identified.

In the sample of angelica (djagilevyj) honey from Altajskiy district the proportion of the pollen was the following: Apiaceae (76.86 %), Asteracerae (10.36 %), Rosaceae (5.87 %) and Brassicaceae (6.91 %).

In the sample of polyfloral honey from Smolenskiy district: Brassicaceae (36.23 %), Polygonaceae (26.69 %), Fabaceae (14.65 %), Rosaceae (8.38 %), Asteraceae (6.84 %), Ranunculaceae (6.45 %), Apiaceae (0.48 %), Boraginaceae (0.28 %).

In the districts under the examine farmers traditionally plant the following agricultural crops which are in the same time milleferous plants and become the base of monofloral honeys: *Helianthus annuus* L., *Fagopyrum esculentum* Moench., *Brassica napus* L., *Onobrychis arenaria* (Kit.) DC.; rare plant – *Phacelia tanacetifolia* Benth., *Galega orientalis* Mill., *Sinapis alba* L. Those crops are usual for many regions and their pollen will be found in honeys from different territories. And only the floristic composition of wild species is important for determination of the honey authenticity. For the honeys from examined districts of Altai Territory the pollen of the following species are typical: *Angelica decurrens* (Ledeb.) B. Fedtsch., *Pastinaca sylvestris* Mill., species of genus *Heracleum* L. etc.; Asteracerae – *Achillea* L., *Bidens* L., *Centaurea* L., *Cichorium intybus* L., *Sonchus* L., *Taraxacum officinale* Weber ex F. H. Wigg. etc.; Caryophyllaceae – *Dianthus* L., *Cerastium* L., *Stellaria* L. etc.; Fabaceae – species of genus *Lathyrus* L., *Medicago* L., *Trifolium* L. etc.; Geraniaceae – *Geranium* L.; Lamiaceae – *Dracocephalum* L., *Origanum vulgare* L., *Thymus* L. etc.; Onagraceae – *Chamaenerion angustifolium* (L.) Scop., species of genus *Epilobium* L.; Rosaceae – *Fragaria* L., *Padus avium* Mill., *Rosa* L., *Rubus* L. etc.; Valerianaceae – species of genus *Valeriana* L.

Among the relicts in studied areas, the potential milleferous have been found. They are *Cimicifuga foetida* L., *Corydalis nobilis* (L.) Pers., *Geranium robertianum* L., *Viola mirabilis* L., *Epilobium montanum* L., *Myosotis krylovii* Serg., *Pulmonaria mollis* Wulf. ex Hornem., *Stachys sylvatica* L., *Galium odoratum* (L.) Scop., *Campanula latifolia* L., *Adonis villosa* Ledeb., *Thalictrum petaloideum* L., *Paeonia hybrida* Pall., *Gymnospermium altaicum* (Pall.) Spach., *Lychnis chalcedonica* L., *Potentilla rupestris* L., *Astragalus glycyphyllos* L., *Scrophularia altaica* Murr., *Alfredia cernua* Cass. etc. In the middle lands and high lands the relicts of tall grasses occurs *Saussurea latifolia* Ledeb., *Allium ledebourianum* Schult. et Schult. fil., *Sanguisorba alpina* Bunge [15].

One of the most important thing for the determination of geographical origin of examined honey samples is the skills of the specialists in identification of pollen "markers" from the local territory, which has small distribution areas. For that purpose, it is needed to form collections of pollen of these plants and to publish catalogue with descriptions and images. In addition, for elaboration of studied honeys it is necessary to compare botanical composition of pollen with other databases of honeys from the other regions.

## References

1. A.N. Burmistrov, V.A. Nikitina, *Honey plants and their pollen* (Moscow, 1990)
2. M.K. Khushnazarov, *Reports of the Academy of Sciences of the Republic of Tajikistan. Botany*, **51**, 10 (2008)
3. R.G. Kurmanov, R. R. Masalimova, A. Ya. Sharipov, Agrarian Russia, (S) (2009)
4. R.G. Kurmanov, A.R. Ishbirdin, *Melissopalinology* (Ufa, 2014)
5. V.V. Progunkov, *Resources of honey plants in the south of the Far East* (Khabarovsk, 2004)
6. V.N. Kulakov, *Honey resources and prospects for the development of beekeeping in the Russian Federation*: *dissertation abstract* (Moscow, 2012)
7. V. von der Ohhe, L. Persano Oddo, M. L. Piana, M. Morlo, P. Martin, Apidologiya, **35** (2004)
8. O. Escuredo, M. Fernández-González, M. C. Seijo, Agriculture, **2** (2012)
9. M.M. Silantyeva, *Conspectus of the flora of the Altai Territory* (Barnaul, 2013)
10. V.I. Vereshchagin, *Honey plants of the Altai Territory* (Barnaul, 1961)
11. *Qualifier to plants of the Altai Territory* (Novosibirsk, 2003)
12. L.V. Kupriyanova, L.A. Alyoshina, *Pollen and spores of plants of the flora of the European part of the USSR*, **1** (Nauka, Leningrad, 1972)
13. *Pollen, Century.* URL: https://www.ages.at/themen/umwelt/pollen/projekt-ponet/ (accessed: 03/19/2019).
14. G.I. Nenasheva, M.S. Ivanova, N.S. Malygina, T. M. Kopytina, Ukrainian Ecological Journal, **7**, 4 (2017)
15. M.M. Silantyeva, *Problems of Botany of South Siberia and Mongolia* (Barnaul, 2008)

# Notes on the taxonomy of *Spiraea aemiliana* and related species (Rosaceae) in Russia

*Vera* Kostikova*, *Victoria* Troshkina

Central Siberian Botanical Garden, SB RAS, Str. Zolotodolinskaya, 101, Novosibirsk, 630090, Russia

**Abstract.** The diversity of the plants of *Spiraea* genus, *Calospira* section in terms of the morphological characters on the basis of 28 natural populations was studied. Two species of *Calospira* section spirea grow in the Asian part of Russia (*S. betulifolia* Pall. and *S. beauverdiana* C.K. Schneid.). *S. betulifolia* Pall. subsp. *aemiliana* (C.K. Schneid.) H. Hara subspecies grows on the Islands in the Asian part of Russia (Sakhalin and the Kuril Islands). Its distinctive features are small round or spatulate (obovate) lamina and the absence of the inflorescence follicles pubescence, as well as a few metrical characteristics that may be auxiliary in determining the taxon.

## Introduction

The research of a polymorphous species complex *S. betulifolia* Pall. – *S. beauverdiana* C.K. Schneid., the plants of *Spiraea* genus, *Calospira* section in Asian part of Russia confirmed the existence of 2 species. *S. beauverdiana* differs from *S. betulifolia* in having dense pubescence of the pedicles and the follicles and the curved position of the follicle rostrum. Two studied species of *Spiraea* genus of *Calospira* section are indistinguishable in terms of quantitative morphological characters. The qualitative composition of the phenolic compounds in aqueous ethanol extractions from spirea leaves is species-specific [1, 2]. Some researchers mark out one more species *S. aemiliana* C.K. Schneid. as a subset of the polymorphous complex *S. betulifolia* – *S. beauverdiana*. This taxon has low height up to 30 cm and round coin-shaped leaves. In terms of the habit it is closest to *S. beauverdiana*. Besides Russia, it also grows in Japan [3]. C.K. Schneider [4] described this species from his research in the high-mountains of Hokkaido, Japan. *S. aemiliana* was described as a separate species by V. N. Voroshilov [5, 6] in the «Flora Sovetskogo Dalnego Vostoka» and the «Opredelitel rasteniy Sovetskogo Dalnego Vostoka». G. Koidzumi [7] ascribed to this species the rank of a variety of *S. betulifolia* Pall. var. *aemiliana* (C.K. Schneid.) Koidz., considering it a synonym of *S. beauverdiana*. The same point of view was expressed by J. Ohwi [8] in the «Flora of Japan» and H. Ikeda [9] in his «Flora of Japan». H. Hara [10] transferred this taxon into the rank of *S. betulifolia* Pall. subsp. *aemiliana* (C.K. Schneid.) H. Hara. V.V. Yakubov [11] noted that on the Southern Kuril Islands one can frequently see a type of *S. betulifolia* with small coin-shaped leaves, which has the appearance similar to the kin species *S. beauverdiana*, differing from it in having completely bare

* Corresponding autor: serebryakova-va@yandex.ru

inflorescence pedicles. In the opinion of N. N. Tsvelev [12], shortgrowing, parvifolious and almost void of pubescence *S. betulifolia* samples from Kunashir Island deserve an independent species' rank which he had denotated as the new *S. vorobjevii* Tzvelev species.

## Materials and Methods

In the Asian part of Russia, *S. betulifolia* has been collected from 17 populations and *S. beauverdiana* from 7 populations in 2003–2017 in Amur, Magadan, Sakhalin and Kamchatka regions, the Jewish Autonomous Oblast, Khabarovsk Krai, Primorski Krai and the Republic of Yakutia. The populations of plants with *S. aemiliana* habit has been collected in two locations on Kunashir Island (the «Kislaya» and the «Kipyashchaya» populations). Each of the population includes not less than 30 specimens. All the collecting was done in July and August, during the follicle ripening period. To study the morphological variability, we selected the qualitative and the quantitative characteristics of the most functionally significant organs: the lamina, blossom clusters and the stalks [13]. The shape of the lamina was determined by the criteria which had been developed earlier [14].

## Results and Discussion

We discovered the plants with the typical small and round shape of the lamina only in the «Kislaya» and the «Kipyashchaya» populations from the Kunashir Island (see the Table). We have not discovered such a lamina anywhere else. The round shape of the lamina prevails in these plants. Also, comparatively small spatulate shape of the lamina has been noted in these populations. The plants from the «Kislaya» and the «Kipyashchaya» populations do not differ from *S. betulifolia* by any other qualitative characteristics. No significant differences in quantitative characters have been discovered. The plants from the «Kislaya» and the «Kipyashchaya» populations have smaller mean values of some metrical characteristics, and they do not overlap with the mean values of *S. betulifolia* and *S. beauverdiana*: «The length of the leaf», «The distance from the leaf base to the widest part of the leaf», «The petiole length», while the relative characteristic «The relative length-to-width ration» and «The ratio of the bud length to the petiole length» are noted more in *S. aemiliana*. However, the scope of the abovementioned characteristics in the plants of the «Kislaya» and the «Kipyashchaya» samples fit with the variability amplitude of *S. betulifolia* and *S. beauverdiana*. The putative *S. aemiliana* plants are closer to *S. beauverdiana* in terms of its quantitative characteristics.

Besides, the genome size of the studied spirea and the chemotaxonomy have been researched. The values of the relative DNA content in the spirea are reliably different with high statistical significance ($p<0{,}001$). The relative DNA content in *S. betulifolia* ($2C = 0{,}91$ and $1{,}01$ pg) is two times higher than the corresponding values for other species. *S. beauverdiana* ($2C = 0{,}55$ and $0{,}5$ pg) and the putative *S. aemiliana* ($2C = 0{,}45$ and $0{,}48$ pg) are more related taxons in terms of DNA content [15]. The study of the interpopulation variability of the *Calospira* spirea in terms of phenolic compounds demonstrated that *S. betulifolia* is different from *S. beauverdiana* in the rutin and avicularine content in water-ethanol extractions from the leaves. Additional components which are absent in *S. betulifolia* have been discovered in *S. beauverdiana* leaves. The putative *S. aemiliana* samples are not much different from *S. betulifolia* sample in terms of the phenolic compounds content, however, the *S. aemiliana* leaves contain isoquercitrin absent in *S. betulifolia* and *S. beauverdiana* [16].

Thus, we should mark out a separate *S. betulifolia* subsp. *aemiliana*, which is rather different in having small round lamina shape on the basis of the research conducted in Russia. This taxon does not differ from *S. betulifolia* in terms of other qualitative characters and is close to *S. beauverdiana* in its quantitative characters.

**Table.** Variability in the shape of lamina of the *Spiraea* genus, section *Calospira*

| Taxon | The shape of lamina | | | | |
|---|---|---|---|---|---|
| | **Ellipsoidal** | **Ooidal** | **Diamond** | **Spatulate** | **Rounded** |
| |  |  |  |  |  |
| *S. betulifolia* | + | + | + | + | – |
| *S. beauverdiana* | + | + | – | – | – |
| «Kislaya» | – | – | – | + | + |
| «Kipyashchaya» | – | – | – | + | + |

We provide the main synonyms for the abovementioned taxons of *Spiraea* genus, *Calospira* section and the key for their determination on the basis of the conducted research.

***Spiraea betulifolia*** Pall. 1784 in Fl. Ross. 1(1): 33; ej 1789 in Fl. Ross. 1(1): 75; Poyark., 1939, in Fl. USSR 9: 288, s. str.; Ohwi, 1965, in Fl. Japan: 521, s. str.; Voroshilov, 1966, in Fl. Sovetsk. Daln. Vost.: 249; Vorobiev, 1968, Dikorastushchiye derev. i kust. Daln. Vost.: 118; Voroshilov, 1982, in Opredel. rast. Sovetsk. Daln. Vost.: 337; Polozhiy, 1988, in Fl. Sib. 8: 14; Yakubov, 1996, Sosud. rast. Sovetsk. Daln. Vost. 8: 134, s. str.; Koropachinsky, Vstovskaya, 2012, Drevesn. rast. Aziatsk. Ros.: 384; Tsvelev, 2008, in Novosti Sist. Vyssh. Rast. 40: 80.

***S. betulifolia*** Pall. subsp. ***aemiliana*** (C.K. Schneid.) H. Hara, 1951, J. Fac. Sci. Univ. Tokyo, Sect. 3, Bot. ser. 3, Bot. 6: 77. ≡ *S. aemiliana* C.K. Schneid. 1905, Bull. Herb. Boiss. Ser. 2. T. 5(4): 347; Voroshilov, 1966, in Fl. Sovetsk. Daln. Vost.: 349; he is, 1982, in Opredel. rast. Sovetsk. Daln. Vost.: 336. ≡ *Spiraea betulifolia* Pall. var. *aemiliana* (C.K. Schneid.) Koidz., 1909, Bot. Mag. (Tokyo) 23: 166; Ohwi, 1965, in Fl. Japan: 521.

**S. *beauverdiana*** C.K. Schneid. 1905, in Bull. Herb. Boiss., 2 ser. 5: 348, incl. var.; Poyark., 1939, in Fl. USSR 9: 289; Voroshilov, 1966, in Fl. Sovetsk. Daln. Vost.: 249; Vorobiev, 1968, Dikorastushchiye derev. i kust. Daln. Vost.: 120; Polozhiy, 1988, in Fl. Sib. 8: 12; Yakubov, 1996, Sosud. rast. Sovetsk. Daln. Vost. 8: 134; Tsvelev, 2008, in Novosti Sist. Vyssh. Rast. 40: 81, hybr.; Koropachinsky, Vstovskaya, 2012, Drevesn. rast. Aziatsk. Ros.: 383. ≡ *S. stevenii* (C.K. Schneid.) Rydb. 1908, in N. Amer. Fl. 22(3): 247; Voroshilov, 1966, in Fl. Sovetsk. Daln. Vost.: 250; Vorobiev, 1968, Dikorastushchiye derev. i kust. Daln. Vost.: 120; Tsvelev, 2008, in Novosti Sist. Vyssh. Rast. 40: 80.

## The key to the taxons of *Calospira* section in Russia

1. Pedicles and follicles are bare, or with scarce or single pubescence, the rostrum of follicle is straight........................................................................................3
2. Pedicles and follicles have dense and wooly pubescence, the rostrum of follicle is curved...........................................................................................................5
3. The stem length is bigger than the bud length, the leaves have ooidal, ellipsoidal,

spatulate or diamond shapes........................................................................*S. betulifolia*
4. The stem length is equal to the bud length, the leaves are small and of round or spatulate shape .......................................................*S. betulifolia* subsp. *aemiliana*
5. The stem equals the bud in length, the leaves are of ellipsoidal or ooidal shape........................................................................................*S. beauverdiana*

## Conclusion

The authors have studied the diversity of the *Calospira* section spirea in terms of the quantitative morphological characters on the basis of 28 natural samples. The key has been made for the identification of species. Two species of *Calospira* section spirea grow in the Asian part of Russia (*S. betulifolia* and *S. beauverdiana)*, which most probably used to be 2 ecotypes of the same species some time ago. *S. betulifolia* subsp. *aemiliana* subspecies grows on the Islands in the Asian part of Russia (Sakhalin and the Kuril Islands). Its distinctive features are small round or spatulate (obovate) lamina and the absence of the inflorescence follicles pubescence, as well as a few metrical characteristics, the average values of which do not overlap with *S. betulifolia* and *S. beauverdiana* average values. This taxon has a trend towards stabilization of morphological characteristics, but has not yet sufficiently separated from *S. betulifolia*.

The reported study was funded by RFBR according to the research (project No. 16-34-00106 мол_а to Kostikova)

## References

1. V.A. Kostikova, T.A. Polyakova, Contemp. Probl. of Ecol., **7**, 3 (2014)
2. V.A. Kostikova, T.A. Polyakova, Bull. of Mosc. Soc. of Natur., 4 (2018)
3. D.P. Vorobiev, *Wild trees and shrubs of the Far East. Determ.* (Nauka, L. 1968)
4. C.K. Schneider, Bull. Herb. Boiss., **5**, 4 (1905)
5. V.N. Voroshilov, *Flora of the Soviet Far East* (Nauka, M., 1966)
6. V.N. Voroshilov, *Determ. of plants of the Soviet Far East* (Nauka, M., 1982)
7. G. Koidzumi, Bot. Mag. Tokio, 23 (1909)
8. J. Ohwi, *Flora of Japan* (DC: Smithsonian Institution, Washington, 1965)
9. H. Ikeda, *Flora of Japan*, **2b** (Tokyo, 1995)
10. H. Hara, J. Fac. Sci. Univ. Tokyo, 3, **6** (1952)
11. V.V. Yakubov, *Vascular plants of the Soviet Far East*, **8** (Nauka, St. Petersburg, 1996)
12. N.N. Tsvelev, Systematics news of higher plants, **40** (2008)
13. T.A. Polyakova *Intraspecific variability of the Far Eastern and Siberian species of the genus Spiraea L .: PhD thesis* (SB RAS, Novosibirsk, 2004)
14. A.A. Fedorov et al. *Atlas on descriptive morphology of higher plants. Sheet* (Edition of the USSR AS, M. – L., 1956).
15. V.A. Kostikova, M.S. Voronkova, E.V. Banaev, T.A. Polyakova, Botan. Pacif., 7(1) (2018)
16. V.A. Kostikova, Mat. of the VI Internat. Scient. Conf. dedicated to the 100th anniversary of Prof. A.V. Polozhiy (State University Publishing House Tomsk, Tomsk, 2017)

# Peculiarities of successional processes taking place in artificial phytocenoses on the Kuzbass coal mining dumps

*Tatiana* Lamanova, *Natalia* Sheremet*, *Vladimir* Doronkin

Central Siberian Botanical Garden, of SB RAS, 630090 Novosibirsk. Russia

**Abstract.** The article presents results of the long-term studies of the structure and production of agrophytocenoses, established on the coal mining dumps in the Kuznetsk basin forest steppe zone. Agrophytocenoses were established on the leveled dumps by seeding grasses and legumes. Another set of control agrophytocenoses was established by the same technic on zonal soil. Seeds were provided by the laboratories of the Central Siberian Botanical Garden SB RAS, Novosibirsk. After 30 yrs of observations it was found that agrophytocenoses on the coal mining dumps did not have initial stage, as well as stage of spontaneous species invasion. Elimination of *Trifolium pratense* and *Arrhenatherum elatius* from coal mining dumps phytocenoses took place 3-4 yrs later than from agrophytocenoses on zonal soils. Species did not eliminate totally, but remained in communities with increased or decreased relative abundances. Significant number of species displayed fluctuation, rather than succession in course of 30 yrs development on the coal mining dumps. These phytocenoses It was also found that agrophytocenoses on the coal mining dumps have high long-term production potential.

## 1 Introduction

Modern researches are focused on vegetation successions due to economic activity [1, 2, 3, 4], and in particular due to open mining of fossil fuel etc. Kuznetsk coal deposit is the biggest one in Russia, both in estimated coal reserves and coal mining. Pit and dump complexes, resulting from the open mining, occupy vast areas in the region, and their revegetation goes slowly. It results in changed regional hydrological regime, contamination of the environment, increased erosion, decreased biodiversity and human health hazard. Restoration of disturbed lands by biological methods is possible by the establishing artificial sylvi- or agrophytocenoses. The latter can be used as agricultural areas, i.e. for hay-making and pastures, as well as recreational areas.

In 1989, by the request of Kuzbassrazresugol concern, we initiated agrophytocenoses establishing on the coal mining dumps, containing bedrock of different geological age, without addition of water-retaining layer and fertile soil layer, without any mineral or organic fertilization, as well as microbial preparations.

* Corresponding author: nsheremet@yandex.ru

## 2 Methods and conditions of the study

The study was carried out on the coal mining dumps of Listvyansky mine (N53.674, E86.961) in the forest-steppe zone 15 km to the south-west of Novokuznetsk, Kemerovo region, Russia.

The local climate is continental with the day temperature sum above 10°C about 1800-1900°C. Annual precipitation is in average 450-600 mm, with a snow cover in winter reaching 30-150 cm. Prior to disturbance by the open mining leached and podzolised chernozems prevailed in the area. Surface mining resulted in the formation of coal mining dumps. Depending on the relative age of the bedrock, we distinguished two groups of dumps. Shallow coal deposits resulted in dumps of the Quaternary sediments, consisting of potentially fertile loesslike loams. Deep coal deposits resulted in dumps of Permian sediments, consisting of argyllites, aleurolites and sands. Granulometrically the sediments can be classified as substrates non-feasible for biological remediation [5].

On the thoroughly leveled dumps, composed by the Quaternary and Permian sediments, of the Listvyansky mine one or several species of grasses and legumes were sawn in July 1989-1990 on 100 ha. Establishment of the sustainable vegetation cover on lands, disturbed by surface mining, is possible due to the large assortment of seed bank and living plants collection, accrued over many years by researchers of the plant introduction laboratories of the Central Siberian Botanical Gardens of the Siberian Branch of the Russian Academy of Sciences. To create sustainable agrophytocenoses on the coal mining dumps we used the following species: common melilot (*Melilotus officinalis* (L.) Pall.), red clover (*Trifolium pratense* L.), hybrid clover (*T. hybrydum* L.), alfalfa (*Medicago sativa* L.), sainfoins (*Onobrychis arenaria* (Kit.) DC.), Cock's-foot (*Dactylis glomerata* L.), crested wheatgrass (*Agropyron cristatum* (L.) Beauv.), smooth brome (*Bromopsis inermis* (Leyss.) Holub), meadow fescue (*Festuca pratensis* Huds.), tall fescue (*F. arundinacea* Schreb.), tall oat-grass (*Arrhenatherum elatius* (L.) J. et C. Presl), Timothy grass (*Phleum pratense* L.), black bentgrass (*Agrostis gigantea* Roth. s. l.).

Conventional geobotanical methods were used to study the agrophytocenoses [6-9]. Starting in the second year of agrophytocenoses' development, every year (1990–2004, 2013 rr.) in July-early August we collected the species-specific aboveground phytomass (4 replicates from each study site) to estimate phytomass production by each plant species and its contribution into the total community production. Overall 428 phytomass samples were collected.

## 3 Results and discussion

Reseearch on vegetation succession is not only theoretically interesting and challenging, but it is also of big practical significance. Succession is defined as a season-independent, directed and continuous sequence of emergence and disappearance of species in a given habitat [10]. According to T.A. Rabotnov [11], successions in plant communities started by sewing mixed seeds on the zonal soil are classified as secondary revegetation successions. Successions on the coal mining dumps, composed of Permian sediments, are classified as primary successions as they are taking place on substrates devoid of soil *per se*. Successions on the coal mining dumps, composed of Quaternary sediments, can be classified as primary ones, but only conditionally, as small amount of roots, rhizomes and seeds can enter the substrate during surface mining.

B.M. Mirkin and T.G. Gorskaya [9] by studying grass mixtures development in different zones of Bashkiriya distinguished three successional stages in fogger crops agrophytocenoses development on zonal soils.

1. The initial stage dominated by explerents. The latter can be weeds or ruderal species. The stage usually lasts for 1-2 years and may not be very well pronounces in case of the high rate of seeds sawn.

2. The stage dominated by cultural crops. N.M. Dainenko and A.R. Matveev [14] revealed that on the zonal soil types this stage begins with elimination of the seeded species and lasts till perennial components are apparently stabilized. The production stage lasts for 1-2 or even 10 years and more.

3. The stage of ruderal species invasion followed by introduction of meadow and steppe species. The plant community is gradually enriched and getting closer in richness to the natural communities.

The longevity of the second stage is determined by the compliance between the edaphic and climatic requirements of the species used and by the risk of introducing some spontaneous species, better than cultural crops adapted to the local environment. According to the ratio of these factors three levels of potential longevity can be distinguished: high (30 and more years), moderate (10 years and more) and low (3-5 years). The moderate longevity was found to be common for mixed grasses and herbs crop.

Our long-term studies of vegetation dynamics in mono- or mixed crops on the coal mining dumps in the Kuzbass region showed that the regularities, commonly observed in artificial phytocenoses on zonal soils, were not always apparent on the coal mining dumps over the study period.

The initial successional stage in the agrophytocenoses was absent. The 1[st] and the 2[nd] years of phytocenoses' development had 80-97% of the aboveground phytomass produced by seeded grasses and legumes over the entire time span of the study. As an example let us consider the tall fescue phytocenoses (Fig. 1):

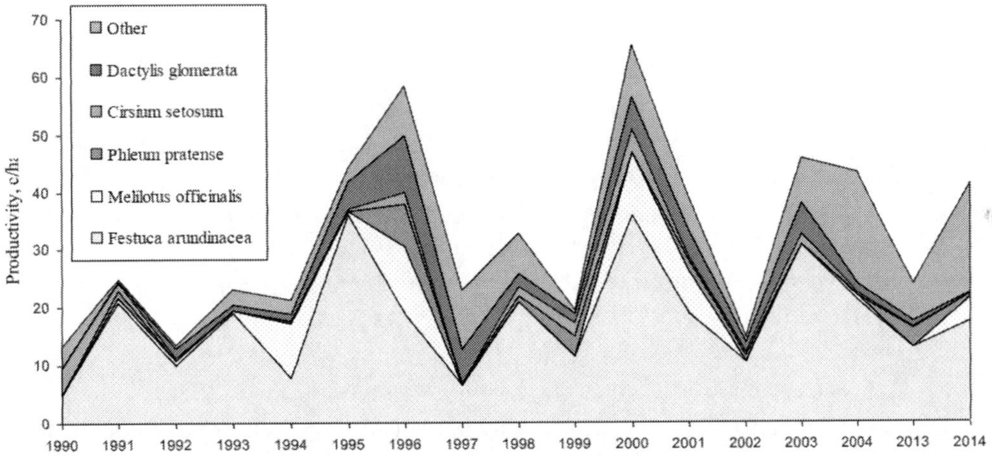


**Fig. 1.** Aboveground phytomass production (centner/ha) and species composition in the *Festuca arundinaceae* phytocenoses on the coal mining dumps in the forest-steppe of the Kuznetsk basin, Russia.

Elimination of such species as *Trifolium pratense* and *Arrhenatherum elatius* from agrophytocenoses grown on the coal mining dumps was found to occur 3-4 years later as compared with the agrophytocenoses, developed on the zonal soils. The species were found to be present during all the time, albeit at increased or decreased relative abundance. Optimal years resulted in increased abundance of the species, upregulating them to codomination [14]. This phenomenon was found to be true for all the species seeded on the coal mining dumps in the Kuznetsk basin [10]. Therefore, the artificial plant communities that we established function as polydominant meadows.

In plant communities initiated by seed mixtures on the zonal soils successional changes can be observed during 10-20 years, whereas in communities, seeded on the coal mining dumps, most of plant species showed only fluctuation changes during 30 and more years. Thus the long-term observations on the structure and production of agrophytocenoses, established on the leveled coal mining dumps in the forest-steppe zone of the Kuznetsk baisin showed that such communities have high potential of productive longevity.

This high productive longevity results from the peculiarities of the ecotope, the correct choice of the plant species to use, the intensity and type of land use, and, most importantly, plant species potentials, our knowledge about which being rather limited. Besides, the possibility for diasporas to enter from external areas is impeded.

Our research showed that alongside the main benefit of low cost of such restoration technology, establishment of artificial phytocenoses on lands disturbed by surface coal mining in the Kuznetsk basin has the following advantages as well:

1) the vegetation cover restoration is possible not only on dumps, composed of Quaternary sediments, but also on dumps composed of Permian sediments, previously believed to be unfavourable for such restoration;

2) starting from 3$^{rd}$ or 4$^{th}$ years of development the plant communities can be used for hay-making and partially for grazing, their productivity equaling or even exceeding the one of the watershed natural meadows existing in the area prior to disturbance;

3) plant communities are characterized by high species richness, that, on the one hand, determined their sustainability, and, on the other hand, is of independent value for the restoration of ecosystem functions and services.

## References

1. A.A. Titljanova, A.D. Sambuu, *Succession in grasslands* (SB RAS, Novosibirsk, 2016)
2. L. Huang, P. Zhang, Y. Hu, Y. Zhao, Global Ecol.Cons., **4** (2015)
3. J.F. Qu, S.L. Zhang, K. Wang, G. Li, N.S. Bu, *Legislation, Technology and Practice of Mine Land Reclamation* (London, 2015)
4. T.Z. Wang, X.D. Huang, Y.Y. Bao, X.R. Gan, *Legislation, Technology and Practice of Mine Land Reclamation* (London, 2015)
5. *State Standard of the USSR*. Protection of nature and lands. Classification of surface mining rocks and substituting substrates for the biological recultivation of lands (GOST 17.5.1.03.-86, 1987)
6. A.A. Korchagin, E.M. Lavrenko, *Field Geobotany* (M.-St.-P., 1976)
7. A.V. Kuminova, *The vegetation of the Altai* (USSR AS Pbs., Novosibirsk, 1960)
8. *Cenopopulations of plants. Main concepts and structure* (M, Nauka Pbs., 1976)
9. R. Whittacker, *Communities and ecosystems* (M, Progress Pbs., 1980)
10. Y.A. Pesenko, *Principles and methods of the quantitative analysis in faunistic studies* (M, Nauka Pbs, 1982)
11. M. Begon, G. Harper, C.R. Townsend, *Ecology: Individuals, Populations, Ecosystems* (M, Mir Pbs., 1989)
12. T.A. Rabotnov, *Experimental phytocenology* (M, Moscow University Pbs., 1987)
13. B.M. Mirkin, T.G. Gorskaya, Biol. Sci., **1** (1989)
14. N.M. Dainenko, N.M. Matveev, *Dynamics of cenopopulations* (M, 1985)
15. T.G. Lamanova, Plant resources, **34**, 3 (1998)

# Flora and vegetation of the Minusinsk town and its vicinity

*Maria* Larina, *Olga* Zyryanova*

Khakas State University named after N.F. Katanov, 90 Lenin St., 655000, Abakan, Russia

**Abstract.** The article demonstrates the results of studies about species composition of lichens, fungi, mosses and higher vascular plants found in the ribbon pine forests in the Minusinsk town and its vicinity. The article based on the original authors' herbaria. 62 basidial macromycetes, 80 lichens and 210 species of the higher vascular plants were found in the studied area. The plant communities and their dominant species were studied.

## 1 Introduction

The town of Minusinsk is an urban district, located in the southern part of the Krasnoyarsk Territory, in the center of the vast forest-steppe Minusinsk Hollow, on the right bank of the Yenisei River [1].

Every year the territory of the town of Minusinsk and its vicinity gets more and more exposed to anthropogenic influence. Forest communities serve as a recreation area for the residents, as well as a dumping ground. Surface fires are often happened in spring and summer. Meadow and steppe communities during the vegetation period are often serves as pastures and places for the wild medicinal plants gathering. These and other factors lead to the changes in the natural flora and vegetation. Therefore, it becomes necessary to study species composition of lichens, fungi, and higher vascular plants in the typical plant communities of the Minusinsk town and its vicinity.

## 2 Material and methods

Authors collected about 550 herbarium specimens of higher vascular plants, 180 lichens, and 60 mushrooms which were the base for the article. The collected herbarium materials were processed in the laboratories of Khakas State University named after N.F. Katanov with the use of binocular MBS-10, microscope "Biolan" and chemical reagents. Species were determined with the help of "Identification Guide to the Plants of the South of the Krasnoyarsk Territory" [2], "Flora of Siberia" [3], "Identification Guide to Lichens of the USSR" [4], "Identification Guide to Lichens of Russia" [5], "Nordic Lichen Flora" [6], "Flora of Fungi of Ukraine" [7, 8], "Identification Guide to Fungi of Russia" [9], «Funga Nordica» [10], «Farbatlas der Basidiomyceten Colour Atlas of Basidiomycetes» [11],

* Corresponding author: o_a_zyryanova@mail.ru

“Floras of Agaric Mushrooms of the Southern Part of the Krasnoyarsk Territory [12], "Mushrooms of the USSR" [13], "Fungi. Identification Atlas" [14].

## 3 Results and discussion

Vegetation cover of Minusinsk town and its vicinity is represented by a variety of different plant communities. Due to the synanthropization and degradation of vegetation cover in urban areas, the urban vegetation is divided into cultivated (park and garden complexes, lawns) and natural (steppes, forests, and meadows) [15]. The main types of green plantings in the town are parks, squares, street greenery, and common lands, while the natural vegetation on the territory of the town is preserved fragmentary and weakly represented in the residential area. The natural plant communities are more often found in the vicinity of Minusinsk, where they occupy large areas. Natural vegetation is comprised of true meadows, less often rocky steppes, valley meadows, forests (pine, birch, mixed birch-pine) and coppices.

The fescue-koeleria low-bunch steppes occupy the southern slopes. The edificators of formation are *Koeleria cristata* (L.) Pers. and *Festuca pseudovina* Hack. ex Wiesb. Abundant species include *Poa stepposa* L., *Agropiron cristatum* (L.) Beauv. and *Cleistogenes squarrosa* (Trin.) Keng.

Among the constant species of low-bunch steppes there are *Stipa capillata* L., *Carex pediformis* C.A. Mey., *C. duriuscula* C.A. Mey., *Medicago falcata* L., *Iris flavissima* L., *Dianthus versicolor* Fisch. ex Link, *Lychnis sibirica* L., *Thalictrum foetidum* L., *Potentilla acaulis* L., *P. bifurca* L., *Goniolimon speciosum* (L.) Boiss., *Schizonepeta multifida* (L.) Briq., *Veronica incana* L., *Galium verum* L., *Scabiosa ochroleuca* L., *Artemisia frigida* Willd. Among the lichens of low-bunch steppes, the constant species include *Cladonia pocillum, Cl. pyxidata* (L.) Hoffm., and a constantly migrating fruticose lichen *Xanthoparmelia camchadalis* (Ach.) Hale. The high species diversity of the low-bunch steppes is complemented with crustose and foliose epilithophytes located on stones. The most frequent crustose epilithophytes are *Aspicilia maculate* (H. Magn.) Oxner, *A. transbaicalica* Oxner, *Lecanora campestris* (Schaer.) Hue, *L. frustulosa* (Dicks.) Ach.,*L. crenulata* Hook., *Candelariella aurella* (Hoffm.) Zahlbr., *Psorotichia schaereri* (A. Massal.) Arnold, *Protoparmeliopsis muralis* (Schreb.) M. Choisy. The most common foliose species include *Xanthoria elegans* (Link) Th. Fr., *Xanthoparmelia somloënsis* (Gyeln.) Hale and *Phaephyscia sciastra* (Ach.) Moberg. The macromycetes are comprised of *Bovista nigrescens* Pers, *Panaeolus papilionaceus* (Bull. ex Fries) Quelet, *Amanita vaginata* (Bull.) Lam.

The sedge low-bunch steppes are confined to the smooth parts of the lower south-exposed slopes, occupying large areas on the plain, including the elevated sites of the river valleys. The main edificator is hardish sedge – *Carex duriuscula*. *Potentilla anserine* L., P. *multifida* L., *Artemisia laciniata* Willd., *Scutellaria scordiifolia* Fisch ex Schrank, *Iris biglumis* Vahl. are often found among the related plants.

The vegetation of stony grasslands or shrubby meadow-steppe is widespread on the tops of bare mountains and southern slopes. In certain places there are thickets of *Caragana pygmaea* or *Spiraea hypericifolia* L., *S. media*, *Cotoneaster melanocarpus*, *Atrophaxis frutescens* L.

The low-bunch meadow-steppes are confined to the smooth southern slopes. Among all steppe communities, the low-bunch true steppes are the poorest in species of lichens and fungi. Typical lichen species are *Xanthoparmelia camschadalis,* X. *somloënsis* (Gyeln.) Hale, *Lecanora frustulosa*, *Aspicilia cinerea* (L.) Körb., *Diplotomma venustum* Körb., while the typical fungi species comprise *Panaeolus campanulatus* (Bull. Ex Fries) Quelet.

Dry meadows occupy the high valley terraces. These meadows include the formations of *Calamagrostis epigeios*, *Trifolium pretense, Sanguisorba officinalis* L., *Amori arepens*, and others [16]. The typical representatives of fungi are *Agaricus campestris* L., *Agrocybe praecox* (Pers.) Favod, *Scleroderma albidum* Pat. & Trab., and others. Due to the lack of competitiveness, the lichens were not found in the meadows.

In the northeast, southeast, and a small area in the southwest of the town, there is Insky ribbon pine forest [17-18] dominated by scotch pine *Pinus silvestris* L. Its shrub layer is formed by *Caragana pygmea* (L.) DC. The herbaceous layer is comprised of *Urtica cannabis, Urtica urens* L., *Fragaria viridis* (Duch.) Weston, *Onobrychis arenaria* (Kit.) DC., *Amoria repens* (L.) C. Presl, *Trifolium pretense* L., *Convolvulus bicuspidatus* Fisch. ex Link, *Lamium album* L., and others. The soil surface is abundantly covered with the moss carpet of *Pleurozium schreberi* (Willd.ex Brid.) Mitt., *Ptilium crista castrensis* (Hedw.) De Not, *Rhytidiadelphus triquetrus* (Hedw.) Warnst. ,*Hylocomium splendens* (Hedw.) Bruch et al., *Polytrichum sp.* The most common species of lichens spread herein include *Cladonia gracilis* (L.) Willd., *Cl. fimbriata* (L.) Fr., *Cl.coniocraea* (Flörke) Spreng., *Cl. arbuscula* (Wallr.) Flot., *Cl. rangiformis* Hoffm., *Evernia mesomorpha* Nyl., *Flavoparmelia caperata* (L.) Hale, *Flavopunctelia soredica* (Nyl.) Hale, *Hypogymnia physodes* (L.) Nyl., *H. tubulosa* (Schaer.) Hav., *Parmelia sulcata* Taylor, *Peltigera canina* (L.) Willd., *Physcia aipolia* (Ehrh. ex Humb.) Fürnr., *Ph. stellaris* (L.) Nyl., and others. The most spread species of fungi are *Lactarius torminosus* Schaeff, *L. deliciosus* (L.) Gray, *Lycoperdon perlatum Pers., L. pyriforme* Schaeff, *Chalciporus piperatus* (Bull.) Bataille, *Psilocybe cubensis* (Earle) Singer, *Psathyrella chondroderma* (Berk & Broome) A.H. Sm., and others. In the southeast (the district of Minusinsk Thermal Power Plant), the stand of *Pinus silvestris* is often mixed with *Betula pendula* Roth. The coppice comprises *Spiraea chamaedrifolia* L., *S. media* Franz Schmidt, *Rosa acicularis* Lindl., *Caragana arborescens* Lam. The herbaceous layer is dominated by *Calamagrostis epigeios* (L.) Roth, with the abundance of *Poa sibirica* Roshev., *Melica altissima* L., *Trisetum sibiricum* Pupr., *Carex macroura* Meinsh., *Iris ruthenica* Ker-Gawl., *Trollius asiaticus* L., *Aquilegia sibirica* Lam., *Thalictrum minus* L., *Vicia unijuga* A. Br., *Rubus saxatilis* L., *Saussurea amara* ( L.) DC., and others. The lichens are represented by the following species: *Cladonia botrytes* (Hagen) Willd., *Cl. carneola* (Fr.) Fr., *Cl. pocillum* (Ach.) Grognot, *Cl. coniocraea* (Flörke) Spreng., *Lecanora symmicta* (Ach.) Ach., *Parmelia sulcata, Hypogymnia tubulosa, H physodes, Flavoparmelia caperata, Flavopunctelia soredica.* Among the macromycetes there are *Fomes fomentarius* (L.) Fr., *Lactarius resimus* (Fr.) Fr., L. *torminosus* Schaeff, *L. deliciosus* (L.) Gray, *Leccinum scabrum* (Bull.) Gray, *Russulavesca* Fr., *Scleroderma albidum* Pat. & Trab., *Clitocybe nebularis* (Batsch.) P. Kumm., *Cyclocybe aegerita* (DC.) Gillet, *Coprinopsis atramentaria* (Bull.) Redhead. Vilgalys & Moncalvo, *Lycoperdon pyriforme* Schaeff, *Suillus luteus* (L.) Gray, *Paxillus involutus* (Batsch.) Fr., and others.

The birch and birch-pine forests are situated nearby the village Malaya-Minusa and mostly comprise the meadow and steppe vegetation. *Betula pendula, Pinus sylvestris* occupy the upper layer, while the lower layer consists of *Caragana arborescens, Spiraea media, Cotoneaster melanocarpus* Fisch. ex Blytt, *Rosa majalis* Herrm., *Crataegus sanguinea* Pall., *Padus avium* Mill., and others. The lichens and fungi are mainly similar to the species of the pine forests.

There are also poplar coppices along the tributaries of the Yenisei River and in the central part of the town. Tree layer mostly consists of *Populus laurifolia* Ledeb., while the herbaceous layer is presented by *Stellaria graminea* L., *S. media* (L.) Vill., *Amoria repens*, and others. There are only 2 species of lichens *Xanthoria parietina* (L.) Th. Fr. and *Caloplaca cerina* Th. Fr. The fungi species include *Pleurotus ostreatus* (Jacq.) P. Kumm. and *Hemipholiota populnea*( Pers.) Bon.

There are also birch-swampy, birch and willow-birch bluegrass-herb-non-swampy coppices found in the valley of the tributary of the Yenisei River.

## 4 Conclusions

The urban ecosystems are characterized by the specific ecological conditions, where the impact of anthropogenic factor manifests itself to the maximum extent [19]. The urbanization process provokes the significant changes both in vegetation and other components of the natural environment. The natural flora is being replaced by the urban flora with populations of alien plant species. Many species of lichens and fungi are not able to adapt to these conditions and consequently die.

## References

1. *Encyclopedia of Krasnoyarsk Krai. South* (Krasnoyarsk, 2008)
2. *Determinant of plants of the South of Krasnoyarsk Krai* (Novosibirsk, 1979)
3. *Flora of Siberia*, **1-14** (Novosibirsk, 1988-2003)
4. *Identification Guide to Lichens of the USSR*, **1-5** (Science, Leningrad, 1971-1978)
5. *Identification Guide to Lichens of Russia*, **6-10** (Science, St. Petersburg, 1996-2008)
6. *Nordic Lichen Flora*, **1-6** (Sweden, 1999-2017)
7. S.P. Vasser, *Flora of Fungi of Ukraine* (Kiev, 1980, 1985)
8. *Flora of Fungi of Ukraine* (Kiev, 2015)
9. M.A. Bondartseva, *Determinant of mushrooms of Russia*, **2** (Science, St. Petersburg, 1998)
10. *Funga Nordica: agaricoid, boletoid and cyphelloid genera* (Copenhagen, 2008)
11. M. Moser, W. Julich, *Farbatlas der Basidiomyceten Colour Atlas of Basidiomycetes* (1988)
12. M.I. Beglyanova, *Flora Agarikovykh of mushrooms of the southern part of Krasnoyarsk Krai*, **1-2** (Krasnoyarsk, 1972-1973)
13. M.V. Gorlenko, *Mushrooms of the USSR* (Moscow, 1980)
14. T.A. Ilyina, *Mushrooms atlas determinant* (Moscow, 2012)
15. O.A. Vasilieva, Ecology of Southern Siberia and adjacent territories, **18 (1)** 2014
16. A.Yu. Korolyuk, N.I. Makunina, Botanical J., **83 (7)** (1998)
17. E.S. Smirnova, Ecology of Southern Siberia and adjacent territories, 19(1) (2015)
18. E.S. Smirnova, O.A Zyryanova, Problems of development of agrarian and industrial complex of Sayano-Altai (2015)
19. A.A. Infantov, Messenger of KRASGAU, **10** (2009)

BIO Web of Conferences **16**, 00017 (2019) https://doi.org/10.1051/bioconf/20191600017
*Results and Prospects of Geobotanical Research in Siberia*

# Coenopopulations of *Myricaria bracteata* Royal at the territory of the Republic of Kazakhstan and their ontogenetic structure

*Elena M.* Lyakh*, *Alexey Yu.* Astashenkov

Central Siberian Botanical Garden, Siberian Branch RAS, 630090 Novosibirsk, Russia

**Abstract.** The article represents the analysis of 4 *Myricaria bracteata* populations in Kazakhstan. It was found that ontogenetic spectrum reflects the development degree of coenopopulations related to succession processes of riparian communities. Depending on the floodplain width, free space and phytocenotic pressure, the ontogenetic spectrum and stability of coenopopulation vary from left-sided incomplete to centered incomplete; from unstable invasive (in a narrow floodplain) through successive transitional and young normal to mature normal stable (in a wide floodplain) ones.

## 1 Introduction

Woody plants play an important role in forming riparian vegetation of floodplains and valleys of mountain and plain rivers. The population approach to assess individual elements of such vegetation occurred to be a justified one, reflecting on one hand the species behavior at different levels of organization (organism and population), on the other hand – various dynamic phenomena taking place under changing ecological and phytocenotic conditions [1]. Such studies make it possible to estimate the level of population stability and plant adaptation, as well as to characterize the succession series associated with both anthropogenic and natural processes in riverside zones.

*Myricaria bracteata* Royal (family Tamaricaceae) is a geoxyl shrub widespread in Central Asia; in the mountains of East Kazakhstan the species is common from Jungar Alatau to Gorny Altai. Its typical habitat is rocky-sandy banks of highland rivers, pebbles, stream dry canals [2]. All species of the genus *Myricaria* are evolving in stream and creek floodplains, and belong to the ecological group of fluviaphytes adapted to short rapid floods. They develop a strong root system and orthotropic whip-like flexible shoots contributing alluvium and sand-silt storage [3]. Growing in floodplain plant groups, *M. bracteata* should be considered as a pioneer tree species providing the upper soil horizon accumulation. At the same time, it can be used as a universal model object to construct succession series reflecting the natural overgrowth of riparian zones. Similar researchhasnot been carried outin Kazakhstan. The work objective is to study the

* Corresponding author: llyakh@rambler.ru

ontogenetic structure of *M. bracteata* cenopopulations (CP) in communities at the primary succession various stages.

## 2 Materials and methods

4 cenopopulations of *M. bracteata* have been studies at the territory of the Republic of Kazakhstan, on floodplains differing topographically as parts of diverse plant communities at the primary succession various stages.

The main population-ontogenetic techniques [4, 5] are used. The ontogenetic spectrum is described based on accounting at least 200 sites by 2 m$^2$ size on transects along and across a riverbed made in a regular way. An individual (genet) and partial formations (ramet) are applied as counting units. The individual ontogenetic states based on qualitative and quantitative morphological features have been described by authors earlier [6]. The population type is determined using classifications by A.A. Uranov and O.V. Smirnova [7] and "delta-omega" by L.A. Zhivotovsky [8]. The following demographic characteristics are used as integral parameters: $\Delta$ – age index [9] and $\omega$ – efficiency index [8]. The ecological density is determined on the number of individuals per a habitable space unit [10].

## 3 Results and discussion

On the assumption of main biological features, *M. bracteata* individuals' ontogenesis is simple, complete, less often complex (as a result of a bush injury and some branches rooting). CP self-maintenance is carried out exclusively by seeds. As it has been found by authors earlier, *M. bracteata* seed production and germination are high: seed productivity coefficient is 65-70 laboratory germination rate is 92-96 % [11, 6].

CP1 was studied in riverbed pebble part of the lower Zhuzagash River bank on the Tarbagatai Range northern macroslope (48.754198 N, 82.385573 E). The floodplain was 100 m width covered with loess clay sediments in some sites. The immature plant community with a total projective cover (TPC) of 10% is formed by single samples of the genus *Salix* ssp. of 30-50 cm height and *M. bracteata* species, the species projective cover (SPC) is 5%.

CP was represented by a locus of 100x500 m size. CP density was 2.07 samples/2 m$^2$ at the study period. The ontogenetic spectrum was left-sided, incomplete (Fig. 1a) with a maximum on virginile individuals (*v*) – 52 % and a large proportion of immature ones (*im*) – 45 %. High values of these plants were related to inspermation, or input by organisms, wind and water of plant seeds or other germs (diaspores) providing infestation [12] and genets transport from the river upper reach. It should be noted that part of virginile individuals were developed due to rooted branches (partial formations of the virginile state), having their own ontogenesis. Low values of young generative groups ($g_1$) 3 % and absence of other ontogenetic groups characterized this CP as unstable invasive at the initial stage of its development. According to "delta-omega" classification CP is young, $\Delta = 0{,}09$; $\omega = 0{,}33$.

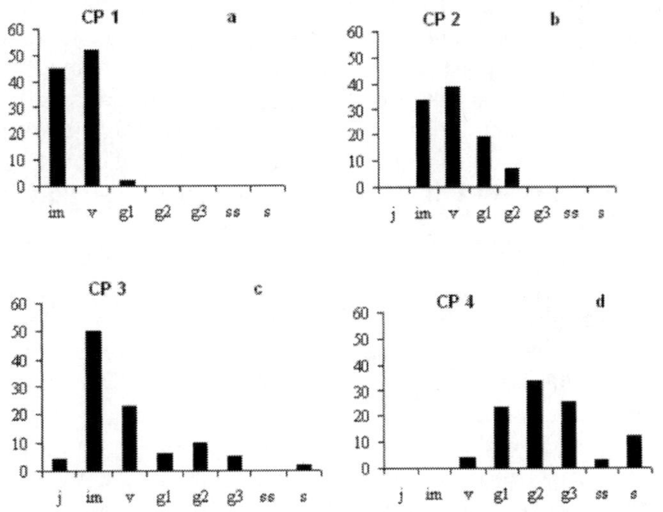


**Fig. 1.** Ontogenetic spectra of *Myricaria bracteata* cenopopulations. X-axis – ontogenetic states; Y-axis – content of ontogenetic groups, %.

CP2 was studied on a large-pebble, sandy-clay bank washed by a temporary channel of the Aksu River (45.638340 N, 79.454040 E). The area of the washed riverine site was 50x50 m. The sparse shrub vegetation group (TPC 60%) was composed by species of the genus *Salix*, *Caragana frutex* (L.) K. Koch, *Myricaria bracteata* 150-200 cm height, SPC of *M. bracteata* – 15 %; *Poa bulbosa* L. dominated in the grass stand consisted of *Barbarea stricta* Andrz., *Thymus marschallianus* Willd. 10-20 cm high.

CP was represented by a locus of 50x50 m size. The individual density was 32.8 samples/2 $m^2$. It consisted mainly of juvenile and immature plants, which were concentrated in the center of the riparian site with accumulated alluvial sandy substrate. Adult generative individuals (*g1* and *g2*) grew along the water level due to limited free space and phytocenotic effect of competitive shrubs (Salix, etc.). The ontogenetic spectrum was left-sided, incomplete (Fig. 1b) with a peak at virginile individuals (39 %). Such spectrum type was similar to CP1spectrum formed as a result of the CP development process. However, the spectrum of CP2 compared to CP1 reflects a trend towards greater stability due to the presence in CP2 of young (*g1*) and middle-aged (*g2*) generative individuals (18 % and 7 %, respectively) ensuring the CP renewal. According to "delta-omega" classification CP is young, $\Delta = 0.16$; $\omega = 0.45$. This CP can be considered as successive one, in a state of transition to normal.

CP3 was studied in bushy riparian floodplain of the Kokterek River on the Tarbagatai Range southern macroslope (46.595213 N, 82.168049 E) on gravels of braided riverbeds on the right bank in a grass-shrub community. The floodplain was 50 m width divided by anthropogenic impact into four ridges coming parallel to the main riverbed and sand spit. The studied community (TPC 70%) was dominated by shrubs: *Salix* spp., *Spiraea hiperecifolia* L., *Caragana frutex*, *Myricaria bracteata* 70-90 cm high, *M. bracteata* SPC-30%; herbaceous species –*Bromus tectorum* L., *Poa bulbosa*, *Stipa zalesskii* Wilensky, 25-30 cm height. This CP occupied an area of 50x100 m and represented a set of five loci: four of them were located along the ridges, the fifth one as on a sand spit. The average density of individuals in the CP was 5.7 samples/2 $m^2$, but it fluctuated sharply at the loci. The CP highest density was at sand spit (74 samples/2 $m^2$), where grew only immature and virginile individuals. In inter-ridge hollows the density ranged from 1.7 to 8.7 samples/2 $m^2$, and was represented by $g_1$, $g_2$ and $g_3$. The CP ontogenetic range was left-sided, incomplete (no specimens of a senile state) (Fig.1 c) with a peak in immature individuals (51 %) reflecting good seed regeneration in previous years and favourable conditions (sand spit) for germination and survival of pregenerative individuals. The gradual decrease in the generative individual portion was associated with phytocenotic pressure by shrubs, which inhibited the development of young plants growing in the inter-ridgehollows. A low portion of post-generative plants was due to the CP development similar to CP2. According to "delta-omega" classification this CP is young, Δ = 0.16; ω = 0.38. It can be characterized as young normal one.

CP4 was studied in the pebble-sandy-clay riverbed part of the Charyn River floodplain, a tributary of the Ili River (43.765111 N, 79.431079 E) in a grass-shrub community. The monodominant shrub community (TPC 45%) is formed by *M. bracteata* of 50–60 cm high. Grasses were represented by *Poa bulbosa*, *Festuca rupicola* Heuff. 20–25 cm high. The floodplain width was 1500 m, the total CP area - 1500x1000 m. CP consisted of 5 loci that remained after automotive vehicles moving along the floodplain. The average density of individuals in the whole CP was 6.3 samples/2 mI. The density ranged from 3.3 to 8.6 samples/2 mI. The ontogenetic spectrum was centered with a peak on middle-aged generative individuals (34%), incomplete (there are no individuals of juvenile and immature states) (Fig. 1d). Pre-generative individuals were confined to sandy loci. But this group was demolished by sand excavation for road construction. The accumulation of mature generative individuals (g3) reflected biological features of the species and was associated with a long generative period, and the floodplain great length (1500 m) effected positively on the population ontogenetic composition and stability. According to "delta-omega" classification this CP is mature, Δ = 0.54; ω = 0.78. The CP can be characterized as stable mature normal one.

## 4 Conclusions

The population analysis of *M. bracteata* in different ecological and phytocenotic conditions has revealed a unidirectional vector of changes in the CP ontogenetic spectrum, which reflects the processes of its development. The trends of the ontogenetic spectrum changesis directed from the left-sided incomplete (with a small set of ontogenetic groups) through the left-sided incomplete (with a large number of ontogenetic groups) to the centered incomplete. Changes in the spectra reflect the species CP succession series while overgrowing riparian zones: unstable invasive → succesive transitive to normal → young normal → mature normal stable ones. *M. bracteata* CP stability is determined by: 1) floodplain width; 2) free space; 3) absence of phytocenotic pressure.

Work is carried out in frames of the projects of the state mission of the Central Siberian Botanical Garden of SB RAS № AAAA-A17-117012610054-6 and № AAAA-A17-117012610053-9.

## References

1. O.V. Smirnova, *The structure of the herbaceous cover of broad-leaved forests* (M, Nauka, 1987)
2. E.G. Bobrov, Bot. J., **52**, 7 (1967)
3. M.T. Mazurenko, Biol. Inland Waters, 3 (2001)
4. A.A. Uranov, T.I. Serebryakova (eds.) *Cenopopulations of plants (basic concepts and structure)* (M, 1976)

5. T.I. Serebryakova, T.G. Sokolov (eds.) *Cenopopulations of plants (essays on population biology)* (M, 1988)
6. E.M. Lyakh, Flora of As. Rus., 2 (2013)
7. A.A. Uranov, O.V. Smirnova, Bull. Moscow Soc. Natur. Explor. Sect. Biol., **74**, 2 (1969)
8. L.A. Zhivotovsky, Ecology, **1** (2001)
9. A.A. Uranov, Sci. Rep. High. Schools. Biol. Sci., 2 (1975)
10. Yu. Odum, *Ecology*, **1** (M, 1986)
11. E.M. Lyakh, Bull. Main Bot. Gar., 4 (2012)
12. V.N. Sukachev, N.V. Dylis (eds.) *Fundamentals of forest biogeocenology* (M, 1964)

# Biological aspects of the genus *Myricaria* Desv. (Tamaricaceae), especially of the Siberian species

*Elena* Lyakh*

Central Siberian Botanical Garden SB RAS, Novosibirsk, Russia

**Abstract**. Observations of the shrub *Myricaria* (Tamaricaceae) in the wild during expeditions and at the Central Siberian Botanical Garden SB RAS (Novosibirsk) has elucidate their taxonomy, and revealed that rather than four taxa, just two are present in Siberia. Without this understanding of the phenology, the systematics of the species was much confused. The two species of Myricaria (*M. bracteata* Royle, *M. longifolia* (Willd) Ehrenb.) in Siberia a are valuable plants for marginal land, urban landscaping and erosion control on river banks. Plants of the genus *Myricaria* are the traditional Tibet medicinal plants. At present, as a result of geological explorative and road-building works, the natural habitats of Myricaria are destroy. The features of blossoming and fructification of *Myricaria* Siberian species of in nature and in cultivation were studied and described for concervation of these valuable plants at the botanical gardens. As a result of the experiments, the temperature features of germination and preservation of seeds was developed.

## 1. Introduction

The genus *Myricaria* Desv. is represented by species distributed in Asia (the Himalayas, southern China, Siberia, and Central Asia) and in Europe (Mediterranean area and Scandinavia). At present time there are about 13 species: Asia, Europe; ten species (four endemic) in China [1].

Two species of *Myricaria* are found in Siberia: *Myricaria bracteata* and *M. longifolia.* They are abundant on short-lived sandbars, gravel banks and other alluvial deposits of the rivers of southern Siberia (Altai, Tuva, West and East Sayan, Buryatia, ets.) [2]. Both species of *Myricaria* are bushes up to 2m in height. The bark of new growth is initially yellowish-green in colour, and become brownish-gray on older stems. The stems bear numerous, densely packed second and third order branches. The leaves are glaucous-green. Their systematic history and morphological variability in Siberia has been reviewed in a publication [3].

Moreover, their deep highly developed roots system makes them excellent for erosion control of vulnerable sites, especially river banks. In addition to the decorative and

* Corresponding author: llyakh@rambler.ru

functional qualities, they are the medicinal plants, honey source, used as fodder and, and produce tannins and flavonoids which have been exploited in the region [4, 5].

At the Central Siberian Botanical Garden extensive cultivation and propagation trials have been undertaken in parallel with a review of the ecology and morphological variability of two species.

## 2. Materials and methods

Observations on the flowering and fruiting of both species was conducted at the introduction nursery of the Central Siberian Botanical Garden, as well as in the wild during expeditions to the Altai, Western and Eastern Sayan. Seeds of *Myricaria* were collected both in the nursery and from wild populations during these expeditions. The research was focused on understanding the conditions under which germination takes place in the natural state; the optimal conditions for germination under controlled laboratory conditions (temperature, humidity, etc.); and recording germinability levels of fresh seed, and seed collected and stored under a number of different conditions

Seed storage and germination requirements were investigated for the purpose of drawing up protocols.

The germinability of *M. bracteata* was investigated by sowing 100 seeds on wet filter paper in a Petri dish. For each treatment, 4 replicates were made. Germination trials were conducted on wet filter paper in Petri dishes in 4 cups- x reconsideration of 100 seeds. The seeds are stored at room temperature (20° C) and at a temperature of 5° C in the refrigerator. We studied the productivity of fruit and seed by the special technique of introduced species [6].

## 3 Results and discussion

The fruit is elongated, pyramidal or pear-shape splits septicidally, and the carpels usually number 3 (2–5). The seeds are small (to 1 mm in length) and bear an apical awn, which has numerous branches in its upper half, allowing the seed to be readily transported by wind. The embryo is large, and straight, and endosperm is absent in the ripe seeds [7].

Earliest signs of germination usually occurs within 8–10 hours of sowing. After 24 hours the seedlings have firmly attached to the substratum, and the cotyledons appear. After 3–4 days, there is a period of intensive root growth, and than this the stem produces the first true leaves on about the 10$^{th}$ day. Under natural situation *Myricaria* seedlings germinate on the sandy and pebbly deposits of river beds where suitable levels of groundwater and surface sediments occur. The seedling are very small, and mist help to dry out. During the first two weeks in particular, the seedlings should be kept almost excessively wet. Usually it is only possible to maintain this with sowings in greenhouses or under plastic film with regular irrigation. However, at the same time this raises the risk of death from excessive heat. Over the course of the first 6 to 8 weeks, until the roots are sufficiently developed, seedlings remain especially sensitive to drought.

After the first growing season, seedlings reach a height of ca. 10–15 cm with 3–5 stems, and possess a well-developed root system. By the end of the second growing season the plants will be 40–45 cm in height, with 7–12 stem. By the third year, growth becomes more rapid, and during May-June, the number of shoots increases to ca. 20, with individual shoots reaching 80 cm in length. Plants grown from seed begin flowering by their 4$^{th}$ year.

In cultivation and in the wild, the shrubs bloom for an extended period of 50–60 days. There are two terms of blossoming and two types of inflorescences: lateral and apical. Lateral inflorescences arise from one-year old wood, while apical inflorescences are borne on the tips of new shoots. The lateral inflorescences begin blooming from the beginning of June to the beginning of July (Fig. A, B), whilst the terminal inflorescences bloom from mid-July to the beginning of August (Fig. C, D).

**Fig. A** - blooming of lateral shoots of *Myricaria bracteata*; **B** - lateral shoots; **C** - blooming of terminal shoots of *Myricaria bracteata*; **D** - terminal shoot.

Observations on *M.bracteata* in culture showed that the lateral flower buds, developed in the previous autumn, from late May to early June, open simultaneously with the opening of new leaves. A single stem bears from 35 to 40 inflorescences, up to a maximum of 60. Mass flowering occurs over a period of ca. 10–20 days. The individual flowers last live about 3 days, following which they fade quickly, and the fruit develops. Developing basipetally, each inflorescence blooms over a period of ca.15–17 days, and extends during this period some 3 to 4 times its original length.

Seed set in high, and the seeds ripen quickly, some 5–6 days after the flower has faded. Individual fruit capsules mature within 14–18 days of the flower fading, they discover how and seeds moving by wind.

Both species are characterized by high fruit and seed productivity, and showed comparable results. Fruit production was compared in natural stands, and amongst cultivated plants at the Central Siberian Botanical Garden.

A wild population of *M. longifolia* was examined in Buryatia, in the Tunkinskaya Valley of the Irkut river. Each lateral inflorescence produced from 15 to 18 fruits, which represenrs 66,5% of the total number of flowers. Non viable seeds determined visually represent approximately 25–30%. Thus a single fruit contains a total of 77–85 viable seed out of a total of 110–120. The figures for plants grown in culture are considerably higher, up to 95%. There is a slight difference in the productivity of stems of first order and second order shoots, which are respectively 60% and 76%. In cultivation such differences did not occur, with the same order of branches giving figure of 94% and 96% respectively.

As the lateral inflorescences finish flowering, herbaceous shoots of the present seasons growth develop apical inflorescences, and these new flower buds are developed, at the Central Siberian Botanical Garden by June 10–18. These shoots are erect, and are typically some 95–120 cm in length. The apical inflorescences typically begin flowering 20–24 June, When the first flowers emerge the inflorescence branches are some 6–9 cm long, lengthen 4-5 times during the course of flowering to some 35–40 in length. The whole compound inflorescence lasts 17–30 days. During this period the central axis may become twisted and may incline downwards. Because new inflorescences are developing and opening all the time, the shrub can be in flower until August 12 to 16 (in the Botanical Garden). New herbaceous shoots develop from May onwards, but this becomes particularly marked after flowering. From the beginning of August to September new flower and leaf buds, that will from the lateral inflorescences and leaves for next year, are developed.

There are no essential differences in flowering behaviour between *M. bracteata* and *M. longifolia. M. longifolia* lateral inflorescences begin blooming from late May to early June, again simultaneous with leaf emergence, and this lasts until mid-June. The apical inflorescences of this latter species have a tendency to be somewhat larger and more complex and less compact.

Valery N. Nekrasov divides the seeds of woody plants according storage into three groups. The seeds of *Myricaria* belong to the first group: they rapidly lose germinability after separation from the maternal plant and require special storage regimes (Nekrasov, 1973) [8]. Two regimes were investigated: storage under room conditions at 20° C, and refrigerated storage at 5° C. These temperatures were selected as they reflect the June temperature regimes of the native mountainous habitats.

Longer-term storage was studied for *M. bracteata* (collected in the Altai Mountains) and *M. longifolia* (collected in Tuva). Seeds stored at room temperature (18–20 ° C) for a month showed complete loss of germinability under the protocols used. Seed stored at a temperature of 3–5° C demonstrated a rapid loss of germinability, to 17–20% after one year. Measurements were undertaken after 1, 2, 6 and 12 month intervals following sowing.

In laboratory conditions, at a temperature of 18–20° C, seeds begin to germinate after 4–6 hours. Seeds, sown on the day of collection, showed germination rates of 27–29% after 8 days. Additional experiments on temperature variation were undertaken for *M. bracteata.* Three parallel regimes were established; firstly seed germinated at room temperature of 20° C; secondly seeds germinated at 30° C; and thirdly seed exposed to 30° C for the 48 hours, and the subsequent days at room temperature of 20° C/ The initial high temperature led to a marked increase in germination rates.

Storage for one night at 5° C, however, resulted in a germination rate of 92–96% after 4 days. Further periods of storage at 5° C for periods from 2 to 5 days, gave similar rates of 93–95%.

## 4 Conclusions

Siberian species of *Myricaria* have two distinct flowering episodes separated by only a few weeks, giving a flowering season up to two months. There are two terms of blossoming and two types of inflorescences: lateral and apical. Without this understanding of the phenology, the systematics of the species was much confused.

*Myricaria* have been shown to have a short viability, and although a large number of seeds are produced each year, the seeds must germinate quickly, preferably after a very short cold period. In the natural conditions of gravel river beds, the cold water constant wetness allow rapid germination to occur. Warm day temperatures also seem to be important for seedling establishment. As a result of geological explorative and road-building works, the natural habitats of Myricaria are destroy. One of possible ways of propagation of Myricaria species is that by seeds.

Seed storage and germination requirements were investigated for the purpose of drawing up protocols. As a result of the experiments, the temperature regime of germination and preservation of seeds was developed.

Seeds have a short germinability, and those of both species should be stored at about 5° C. Germination rates for fresh, viable seeds is 90%, and declines to 20% after 12 months storage at 5° C. Laboratory germination rates doubled when seeds were subjected to an alternating temperature regime (30° C followed by 20° C).

## Acknowledgements

I would like to thank Dr. Matthew Jebb, Director of the National Botanical Gardens of Ireland for value advices and for assistance in translating this publication.

Work is carried out in frames of the projects of the state mission of the Central Siberian Botanical Garden of SB RAS № AAAA-A17-117012610054-6 with applying of bioresource scientific "Collection of Living Plants in Open and Closed Ground" USU 440534.

## References

1. Z.Y. Wu , P.H. Raven, D.Y. Hong, (eds.), *Flora of China*, **13** (Beijing, 2007)
2. G.A. Peschkova (ed.), *Flora of Siberia*, **10** (Science Publishers, Enfield, NH, USA, 2006)
3. E.M. Lyakh, Eurasian Journal of Forest Research, 16 (2013)
4. A.A. Chernonosov, E.A. Karpova, E.M. Lyakh, Revista Brasileira de Farmacognosia, 5 (2017)
5. E.M. Lyakh, N.V. Tsybulya, Plant Resourses, **45**, 1 (2009)
6. M.G. Nikolaev, I.V. Lianguzova, L.M. Pozdova, *Biology of Seeds* (SPb., Nauka, 1999)
7. A.L. Takhtajan, *Comparative anatomy of seed,* **4** (SPb., Nauka, 1992)
8. V.I. Nekrasov, *Basics of Seed Science of Wood Plants in Introduction* (Moscow, Nauka, 1973)

# Epiphytic lichens of the dark coniferous forest within the outbreak of Ips sexdentatus Boern. in the Maly Abakan site of the Khakassky Nature Reserve

*Evgenia* Makeeva[1,2*], *Olga* Zyryanova[1]

[1]Khakas State University named after N.F. Katanov, 90 Lenin St., 655000 Abakan, Russia
[2]State Nature Reserve «Khakasskii», 164 Tsukanova St., Abakan, Russia

**Abstract.** The article demonstrates the results of studies epiphytic lichens species composition from the territory of the Khakassky State Nature Reserve, conducted in the alpine cedar taiga at the location of the *Ips sexdentatus* outbreak. There were two types of sample plots: damaged plots, characterized by shriveled cedars infected by bark beetle and flawless plots containing forest stand of primarily vigorous trees. All sample plots were found to comprise 46 species of lichens belonging to 24 genera and 13 families. It was observed that plots infected by bark beetle were inhabited by lichens with poor vital activities manifested by their damaged thalli. However, there were no evident changes in the epiphytic lichen communities on floristic and coenotic levels.

## 1. Introduction

Epiphytic lichens are one of the components of the boreal forest communities. The characteristics of the epiphytic lichen cover are closely related to certain environmental factors and naturally change in accordance with the changes in environmental conditions. One of such factors, for instance, may be density and length of tree crowns which serve as samples for descriptions [1].

The Maly Abakan site is one of the cluster sites of the Khakassky State Nature Reserve, located in the mid-mountain and high-mountain parts of the Western Sayan northern macroslope. The mountain-taiga belt occupies 80% of the site territory. The most part of the site is represented by taiga dark coniferous forests composed of cedars and firs primarily forming the mixed stands [2].

In recent decades the Maly Abakan site has been found to comprise the shriveling Siberian cedar tree stands exposed to Ips sexdentatus Boern. Besides, the state of cedar forests has been monitored [3, 4]. The years of the maximum distribution of bark beetles on the site are 2000 and 2010. Therefore, the aim of this study is to identify species composition of the epiphytic lichen cover at the location of the Ips sexdentatus outbreak on the Maly Abakan site of the Khakassky Nature Reserve.

* Corresponding author: meg77@yandex.ru

## 2. Material and methods

The species composition of epiphytic lichens was studied at four permanent sample plots (20x20 m in size) located in two study areas. The first plot – Nizhny is located in the vicinity of Nizhny cordon. This plot included sample plots Nd (Nizhny damaged) and Nf (Nizhny flawless). The plots are located on the second terrace of the Maly Abakan River at altitude of 730 m above sea level. The plots are characterized with cedar-fir and moss-forb forests. The coordinates of the sample plot Nd are 52°11'07.4" N; 089°03'36.6" E. According to observations there was an active damage in the cedar stand with absence of vigorous trees, while the old-dried trees accounted for 72%. The coordinates of the sample plot of Nf are 52°11'09.4" N; 089°03'41.4" E. The forest stand of the plot was comprised of 80.2 % of the totally vigorous and slightly weakened cedar trees [5].

The second plot (Otkyl) is located in the mouth of the Otkyl River. There were sample plots Od (Otkyl damaged) and Of (Otkyl flawless) located at altitude of 790 m above sea level. The plot is characterized by cedar-fir and gramineous-forb forests. The sample plot Od is located on the first terrace of the Maly Abakan River with the following coordinates 52°07'21.2" N; 089°04'08.8" E. The plot was comprised of 79% of the old dead wood in the total composition of the forest stand. The sample plot Of is located on mountain slope with following coordinates: 52°08'04.2" N; 089°04'35.5" E. The forest stand of the plot was composed of 90 % of the vigorous and slightly weakened trees. 10 phorophytic trees *Pinus sibirica* Du Tour (5), *Abies sibirica* Ledeb. (5) from each plot were selected for studies. The lichens were collected from the trunks at four heights: 0.6, 0.9, 1.2 and 1.5 m [6].

## 3. Results and discussion

Two studied sites were found to comprise 46 species of lichens belonging to 24 genera and 13 families. An average value of species in the family is 3.5 %; while the following families identified as dominant in the studied flora have the values exceeding this indicator: *Parmeliaceae* Zenker (24 species, 52.8%) and *Cladoniaceae* Zenker (6 species, 13.2 %). The single-species families are *Alectoriaceae* (Hue) Tomas, *Coniocybaceae* Reichenb, *Lecanoraceae* Körb., *Lobariaceae* Chevall., *Mycoblastaceae* Hafellner, *Bacidiaceae* W. Watson, *Mycobilimbiaceae* Hafellner.

An average value of species in genus is 1.92 %. The dominant lichen genera in studied area are *Cladonia* Hill. ex P. Browne (6 species, 13.2 %), *Bryoria* Brodo et D. Hawksw. (6 species, 13.2 %), *Usnea* Dill. ex Adans. (4 species, 8.8 %), *Hypogymnia* (Nyl.) Nyl. (3 species, 6.6 %).

Based on the classification by N. V. Sedelnikova [7], in terms of relation to humidity, snow cover thickness and temperature factor there are 2 ecological groups among the studied epiphytic lichens. They are mesophytes (45 species, 97.8 %) and xeromesophytes (1 species, 2.2 %). This ratio of dominant families, genera, and ecological groups reflects peculiarities of the boreal floras of the Holarctic [8, 9].

Geographical analysis of epiphytic lichens in studied sites demonstrated predominance of the boreal geographical element (22 species, 48.4 %) together with nemoral (8 species, 17.6 %) and mountain (6 species, 13.2 %) elements. Analyzing the type of habitat one can say that the holarctic (17 species, 37.4 %) and pluriregional (13 species, 28.6 %) lichens are predominant.

Consider the species composition of lichens on different sites and sample plots. The total number of lichens found on the damaged and flawless plots nearby the Nizhny cordon is 39 species, whilst the lichens found in the mouth of the Otkyl River account for 23 species. The Jaccard similarity coefficient ($K_J$) for the two studied sites is 0.35, which indicates rather weak similarity of the species composition.

Common species for the two studied plots were *Chaenotheca ferruginea* (Turner ex Sm.) Mig., *Hypogymnia physodes* (L.) Nyl., *H. vittata* (Ach.) Parrique, *Lecanora carpinea* (L.) Vain., *Lobaria pulmonaria* (L.) Hoffm., *Nephromopsis laureri* (Kremp.) Kurok., *Parmelia sulcata* Taylor, *Pertusaria amara* (Ach.) Nyl., *Physcia adsendens* (Fr.) H. Olivier, *Platismatia glauca* (L.) W. L. Culb. et C. F. Culb., *Ramalina dilacerata* (Hoffm.) Hoffm., *Tuckemanopsis chlorophylla* (Willd.) Hale, *Usnea dasopoga* (Ach.) Nyl., *U. glabrescens* (Nyl. ex Vain.) Vain., *U. longissima* Ach., *Vulpicida pinastri* (Scop.) J.-E. Mattsson et M. J. Lai.

Specific species for the plot in the mouth of the Otkyl River included *Bryoria simplicior* (Vain.) Brodo et D. Hawksw., *Cladonia bacilliformis* (Nyl.) Glück, *Cl. gracilis* (L.) Willd., *Cl. rei* Schaer., *Cl. squamosa* Hoffm., *Lecidea turgidula* Fr., *Ramalina pollinaria* (Westr.) Ach.

Specific species for the plot located in the vicinity of the Nizhny cordon consisted of *Alectoria sarmentosa* (Ach.) Ach., *Bacidia circumspecta* (Nyl. ex Vain.) Malme, *Bryoria capillaris* (Ach.) Brodo et D. Hawksw, *B. furcellata* (Fr.) Brodo et D. Hawksw., *B. fuscescens* (Gyeln.) Brodo et D. Hawksw., *B. implexa* (Hoffm.) Brodo et D. Hawksw., *B. nadvornikiana* (Gyeln.) Brodo et D. Hawksw., *B. tortuosa* (G. Merr.) Brodo et D. Hawksw., *Cetrelia cetrarioides* (Delise et Duby) W. L. Culb. et C. F. Culb., *Cladonia coniocraea* (Flörke) Spreng., *Cl. fimbriata* (L.) Fr, *Evernia mesomorpha* Nyl., *E. esorediosa* (Müll. Arg.) Du Rietz, *Hypogymnia bitteri* (Lynge) Ahti, *Lecania cyrtella* (Ach.) Th. Fr., *Lecidea albofuscescens* Nyl., *Melanelia exasperata* (De Not.) Essl., *Mycobilimbia hypnorum* (Lib.) Kalb et Hafellner, *Mycoblastus affinis* (Schaer.) T. Schauer, *Parmelia squarrosa* Hale, *Pertusaria flavida* (DC.) J. R. Laundon, *Physcia tenella* (Scop.) DC., *Usnea hirta* (L.) Weber ex F. H. Wigg.

The damaged and flawless sample plots of the studied sites had the similar number of species. The number of lichens species found on the plots can be described as follows: $N_d$ plot – 25 species, $N_f$ – 28 species, $O_d$ and $O_f$ – 15 species each.

*Pinus sibirica* of the Nizhny site was found to comprise 18 species of lichens on each of $N_d$ and $N_f$ plots; 11 species on the $O_d$ plot and 13 species on the $O_f$ plot. Consequently, the number of lichen species on the sample plots is essentially independent of the degree of damages observed in *Pinus* bark.

Epiphytic species on *Pinus sibirica* found only on the flawless plot $N_f$ include *Bacidia circumspecta, Cetrelia cetrarioides, Hypogymnia bitteri, Lecania cyrtella, Mycobilimbia hypnorum, Mycoblastus affinis, Parmelia squarrosa, Platismatia glauca, Ramalina dilacerata*; on the flawless plot $O_f$: *Bryoria simplicior, Cladonia gracilis, Hypogymnia physodes, Lecidea turgidula, Lobaria pulmonaria, Platismatia glauca, Ramalina dilacerata, Tuckermanopsis chlorophylla*.

The plots infected by bark beetle were found to have some species which were absent on the flawless plots. These species included *Bryoria implexa, B. nadvornikiana, Chaenotheca ferruginea, Evernia esorediosa, Melanelia exasperata, Pertusaria flavida, Physcia adscendens, P. tenella, Vulpicida pinastri* located on *Pinus sibirica* of the plot $N_d$; and *Cladonia rei, Nephromopsis laureri, Parmelia sulcata, Pertusaria amara, Physcia adscendens, Usnea longissima* located on the plot $O_d$.

In contrast to *Pinus sibirica, Abies sibirica* is a forophyte with a smooth and undamaged bark. Species compositions of epiphytic lichens on *Abies* can only be influenced by the changes in light environment in biotopes with shrivelling *Pinus*. *Abies sibirica* was found to have 27 species of lichens in the Nizhny site and 17 species in the Otkyl site. 17 species of lichens were found on the plot $N_p$ and 20 species on the plot $N_f$, of the Nizhny site, while the Otkyl site comprised 9 species on the plot $O_d$ and 10 species on the plot $O_f$. The total number of species found on *Pinus* (on the area of all plots) was 34, on *Abies* – 32.

When comparing lichen biota of all the plots affected by the bark beetle and the flawless plots, only a few species were found to be reducing in number on the damaged biotopes. They were *Platismatia glauca* (reducing from 25.0 % on the flawless plots to 10.0 % on the damaged plots), *Hypogymnia physodes* (from 22.5 % to 10.0 %), *Vulpicida pinastri* (from 17.5 % to 5.0 %).

In case when trees are damaged by bark beetles, the properties of the vegetation substrate are also changed which can directly affect the expansion of lichens. The studied plots demonstrated changes in the morphology of thalli in lichens living on the trees damaged by bark beetles. The necrosis of blastema was mostly observed in foliose (*Lobaria pulmonaria, Parmelia sulcata, Cetrelia cetrarioides*) and fruticose (*Cladonia coniocraea, Cladonia fimbriata*) epiphytic lichens. Moreover, their colour, number of reproductive structures and soredial properties of thalli had undergone certain changes.

## 4. Conclusion

Two selected sites initially differed from each other in species composition, the largest number of species was found on the Nizhny site. The number of lichen species in the areas affected by the bark beetle does not serve as marker of inhibition of epiphytic communities, as the damaged and intact sample plots are quite comparable in terms of the species number. Besides, the part of the species was found only on the flawless plots, as well as the other part was characteristic only for the damaged plots. In our case, it is the damage of the thalli that indicates the inhibition of the vital activity of lichens in the areas affected by the bark beetle, even in the absence of changes in the epiphytic lichen communities at the floristic and coenotic levels.

This work was supported by the RFFI grant 18-44-190004 r_a Assessment of the state of pine forests of the Republic of Khakassia in the place of localization of breeding outbreak of reproduction of secondary stem pest Ips sexdentatus (on the example of the Maly Abakan natural boundary, Khakassky Reserve).

## References

1. *Methods of studying of forest communities* (NIIHimiya SPbSU, St. Petersburg, 2002)
2. *Khakassky Reserve: a scientific publication* (Journalist, Abakan, 2001)
3. I.L. Maymanakova, Scientific research of reserves and national parks of Southern Siberia, **1** (2011)
4. I.L. Maymanakova, Scientific research of reserves and national parks of Southern Siberia, **4** (2014)
5. I.L. Isaeva, Achievements of modern natural sciences, **12 (1)** (2018)
6. A.V. Pchelkin, V.B. Slepov, *The use of algae and lichens in environmental monitoring and bioindication studies* (MGSUN, Moscow, 2004)
7. N.V. Sedelnikova, *Lichens of Altai and Kuznetsk highlands: Summary of flora* (Science, Novosibirsk, 1990)
8. N.S. Golubkova, *Analysis of lichen flora of Mongolia* (L., Science, 1983)
9. O.A. Zyryanova, In the world of discoveries. Series: Natural and technical science, **12.1 (60)** (2014)

# On the validation of spatial statistical vegetation models

*Igor* Makhatkov*

Institute of Soil Science and Agrochemistry SB RAS, Academician Lavrentiev Ave., 8, building 2, Novosibirsk, 630090, Russia

**Abstract**. The features of coefficients of determination and coefficients of leave-one-out method for spatial vegetation model and spatial models of squared deviations are discussed. The properties of models are illustrated in key area for spatial model of Cladonia stellaris projective cover.

Spatial modeling of the vegetation and its different features can be realised as spatial extrapolation of classification units of vegetation [1] or spatial extrapolation of unclassified variables [2]. The last way makes opportunity to create a spatial model of unclassified information [3]. In both cases, spatially distributed variables e.g. the spectral responses of the surface are used as predictors of extrapolation. The quality of modeling is estimated by the coefficient of determination ($R^2$, r-squared), which shows the ratio of variation (the sum of the squared deviations) explained by the adopted model [4]. This estimation of the statistical dependence of modified variables and predictors can be increased unlimitedly by increasing the flexibility of the regression functions and by including more predictors. The validity of such a model improvement is evaluated by validation methods, such as the coefficient of determination based on the leave-one-out method, the successive exclusion of each sample from the training data and the use of its deviation from the predicted value [5]. Depending on the completeness of the data and the features of the regression model, its deviations from the observed values in different areas of the model can vary significantly. Spatial modeling of these deviations can provide a detailed assessment of the quality of the model and the quality of area survey. To investigate deviations of spatial models we used the key area with geobotanical relevés. Spatial models of the squared deviations of the vegetation model and its individual variables from the observed data, and the squared deviations of the leave-one-out method are considered.

25 geobotanical relevés of the key area (N 63.379, E 75.865 / N 63.105, E 76.451), Yamal-Nenets Autonomous Region were used as observed data. 88 rare species were excluded. The key area is within the subzone of the northern taiga [6]. Here the good drained loamy habitats are occupied by zonal larch shrub-moss forests, drained sands are occupied by intrazonal pine shrub-lichen forests. Flat watersheds are occupied by frozen and thawed mires. Oligomesotrophic bogs occupy hydrological net of swamps. Floodplain series of vegetation are common near rivers and streams. Forests and frozen mires on the key area have been partly burned by recent fires. The spatial modeling of species abundance was based on the variables of the spectral responses of the composite of 6

* Corresponding author: makhatkov@mail.ru

channels of 3 summer Landsat images with a resolution of 30 m/pixel. For regression analysis, polynomial functions were used. For data handling we used QGis 2.12 (http://www.qgis.org/) with additional modules, for calculations - Python 2.7 (https://www.python.org/) with the necessary libraries (http: // www. numpy.org/, http://scikit-learn.org/, http://www.gdal.org/), and the editor of PyCharm Community (https://www.jetbrains.com).

The vegetation model of the key area was build using principal components values of the geobotanical relevés matrix with values of species projective cover (%) as factors [7]. According to a preliminary estimation of the principal components significance the first three components were significant only. The coefficients of determination were calculated for the vegetation model in total and separately for 3 factors (main components). In addition, as an example, the coefficients for the projective cover models of two species of Cladonia stellaris (Opiz) Pouzar et Vĕzda and Rubus chamaemorus L werelculated (table).

**Table.** Coefficients of determination

| | Model in total | PCA components | | | Cladonia stellaris | Rubus chamaemorus |
|---|---|---|---|---|---|---|
| | | 1 | 2 | 3 | | |
| Model of vegetation cover | 0.66 | 0.63 | 0.82 | 0.53 | 0.70 | 0.35 |
| Leave-one-out method for model of vegetation cover | 0.52 | 0.52 | 0.72 | 0.32 | 0.63 | 0.13 |
| Model of squared deviations | 0.26 | 0.28 | 0.17 | 0.30 | 0.16 | 0.30 |
| Leave-one-out method for model of squared deviations | 0.40 | 0.38 | 0.43 | 0.43 | 0.19 | 0.54 |

The coefficients of determination of the predicted and observed values were relatively low. This indicates a weak statistical dependence thematic values and predictors, in our case with species abundance and imagine values. Nevertheless some regularities of this dependences should be noted. The models of deviations in all cases were than better than models of vegetation worse. In other words the larger the ratios of deviations that the accepted model does not explain, the more closely these deviations are connected with predictors. This property of deviations is probably associated with differences in the regression functions accepted in the main model building and in model of deviations, and can be used to improving the models.

The greatest deviations of the leave-one-out method are associated with local lack in the survey data when the exclusion of some observations strongly changes the model. The coefficients of determination of leave-one-out method are always better than coefficients of models, and the statistical relationship of deviations and predictors does not always depend on the quality of the model. The values of squared deviations of leave-one-out method can be used as an assessment of the surveying of the territory and pay attention to areas with relatively large values of squared deviations to supplement the survey data.

The properties of deviation models are well illustrated by their fragments for Cladonia stellaris for key area (Fig., a). The projective cover model (Fig., b) shows the highest values for pine shrub-lichen forests that are common on sandy habitats along some rivers and low values for thawed mires, the lowest for larch forests, and complete absence on floodplains. The model of deviations (Fig., c) shows the highest quality of abundance model for most of the larch forests and frozen mires, slightly less for pine forests, and the worst for ones impacted by fires and partly for larch forests. The model of squared deviations of leave-one-out method (Fig., d) is generally similar to the previous one. But it shows large differences on fragments of frozen bogs that were affected by fire. This indicates a lack of survey data in similar areas.

In general despite the relatively weak dependence between squared deviations of model, squared deviations of leave-one-out method and predictors, spatial models their values can be used for a detailed assessment of the quality of the main model and its improvement.

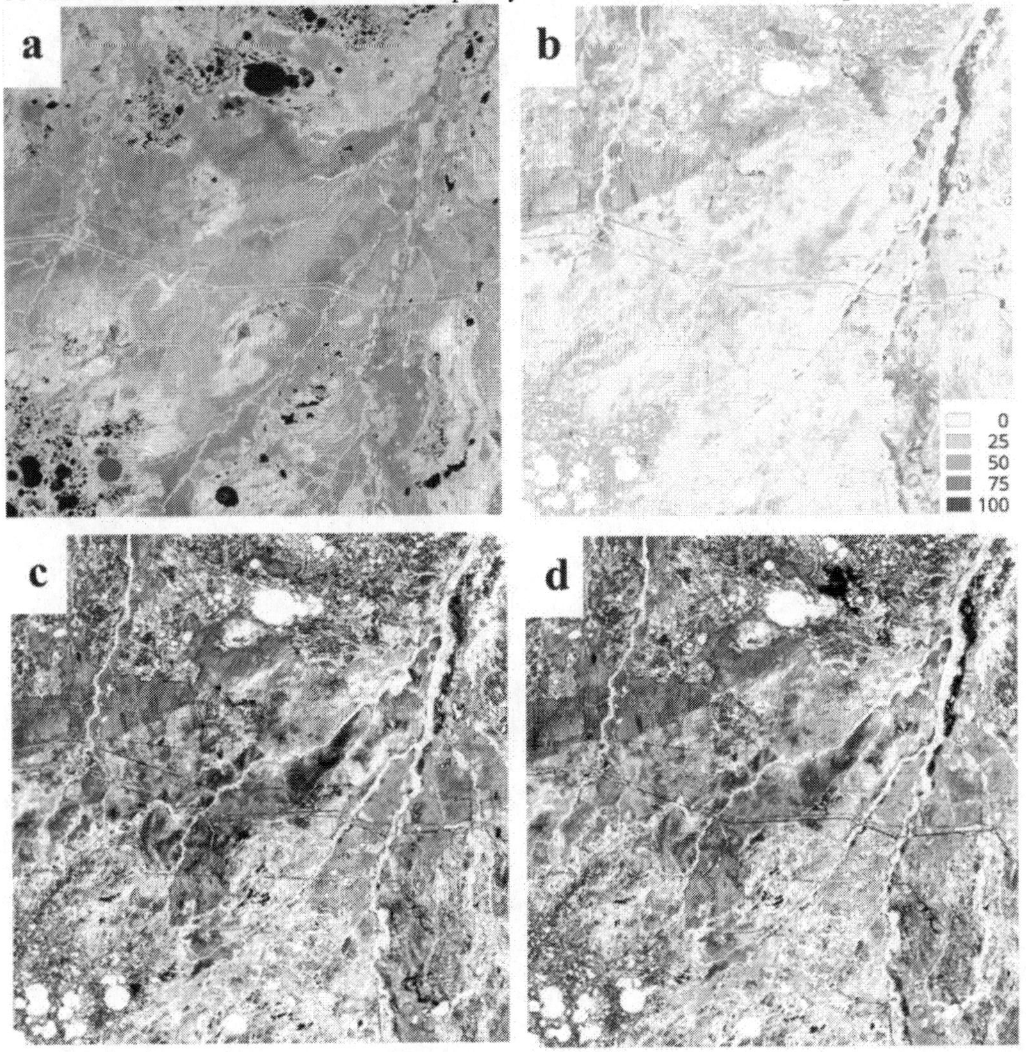


**Fig. a-d.** The spatial model of the projective cover of Cladonia stellaris and the assessment of its reliability (a - image of a fragment of the key site, 7 4 2 Landsat bands, b - model of projective cover (%), c - model of squared deviations, d - model of squared deviations of leave-one-out method).

## References

1. S. Bartalev, V. Egorov., V. Zharko, E. Lupyan, D. Plotnikov, S. Khvostikov, Modern problems of remote sensing of the Earth from space, **12**, 5 (2015)
2. M. Al-Hamdan., J. Cruise, D. Rickman, D. Quattrochi, Remote Sensing, **6 (10)** (2014)
3. J. Franklin, *Mapping Species Distributions: Spatial Inference and Prediction* (Cambridge University Press, Cambridge, UK, 2009)

4. G. Ivchenko, Yu. Medvedev. *Introduction to mathematical statistics* (M., Publishing House of LCI, 2010)
5. B. Efron. *Non-traditional methods of multivariate statistical analysis. Finance and Statistics* (M., 1988)
6. I.S. Ilyina, E. I. Lapshina, N. N. Lavrenko, L. I. Mel'tser, E. A. Romanova, B. A. Bogoyavlenskii, and V. D. Makhno, *Plant cover of the West Siberian Plain* (Nauka, Siberian Branch, Novosibirsk, 1985)
7. J. Kim, C. Muller. *Factor Analysis: Statistical Methods and Practical Issues. Factor, discriminant and cluster analysis* (M., Finance and Statistics, 1989)

# On plant communities of arid regions of Altai-Sayan mountains

*Natalia* Makunina*, *Mikhail* Telyatnikov, and *Evgeniy* Zibzeev

Central Siberian Botanical Garden, 630090 Novosibirsk, Russia

**Abstract.** 8 basic types of plant communities of arid regions of Altai-Sayan mountains have been revealed, their floristic classification has been carried out. Short description of the basic types has been given, their altitudinal distribution has been characterized.

## 1. Introduction

The South-East Altai and the South-West Tuva (Fig. 1) form the northern outpost of Central Asian arid area, while the main part of Altai-Sayan mountains belongs to the boreal zone. The South-East Altai center is the largest Altaian depression – the Chuya depression (1800–2100 m); the steep flanks of mountain ranges encircle the depression. Kuraisky ridge (2800–3800 m) forms its northern border, the spurs of North Chuysky ridge (2500–3000 m) – the western, and Chikhachev ridge (3000–4000 m) – the eastern, Ukok plateau and the Sailugem ridge (2900–3200 m), – the southern. Rivers of the South-East Altai originate in nearby ridges and flow into Chuya river [1]. The South-West Tuva includes Mongun-Taiga mountains (3400-3900 m), the southwestern flank of Tsagan-Shibetu ridge (3300-3500 m) and Kargy depression (1800-2100 m). Shallow mountain rivulets flow into Kargy river.

Three main types of climate appear to correspond to three main geomorphological units: (1) the climate of Chuya and Kargy depressions, (2) the climate of the flanks of the mountain ridges encircling depressions and (3) the climate of high-altitude ridges and plateaus. Low winter temperatures are caused by Siberian anticyclone; the cold air flows down from the slopes, fills the intermountain depressions and creates temperature inversions (January temperature in Chuya depression is –32 °C). Mountain slopes are noticeably warmer than depressions: the average January temperature is –16°C. The climatic conditions of mountain flat tops are similar to those in a free atmosphere [2]: the average January temperature at altitude of 3000 m is –18°C. In the summer, the Chuya depression is the warmest place in the South-East Altai: the average temperature in July is + 14°C, while on the surrounding mountains slopes it is +11°C. At an altitude of 2500 m, the average July temperature goes down to +8°C, and at an altitude of 3000 m – to + 4°C. Annual rainfall in depressions varies from 100 to 150 mm. The surrounding mountain slopes receive up to 400 mm of precipitation; annual rainfall at the height of the snow line reaches 600 mm on the Chikhacheva ridge and 800–1000 mm on the South Chuysky ridge [2].

* Corresponding author: natali.makunina@mail.ru

The altitudinal zone series is known to comprise two belts (altitudinal zones): steppe (1800–2400 m) and alpine (2400–2800 m) one. The limits of altitudinal zones correspond to the main relief forms [3].

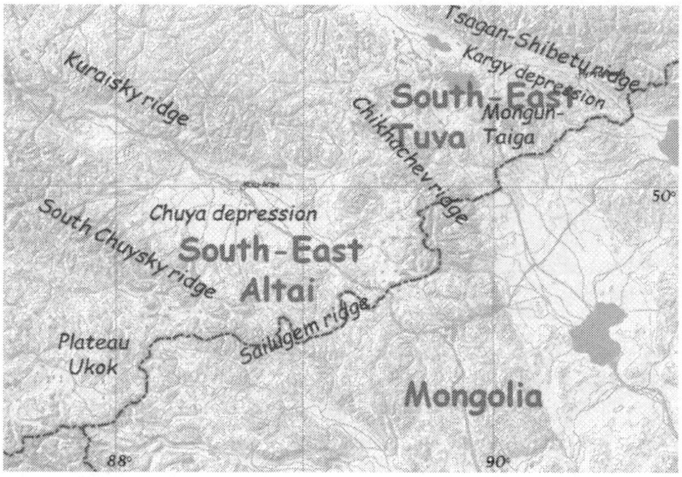


**Fig. 1.** Geographycal position of arid regions of Altai-Sayan mountains.

## 2 Results

### 1.1 Biodiversity

In arid regions of Altai-Sayan mountains we revealed 8 basic types of plant communities: i – shrub tundra, ii – cryophytic forest, iii – alpine meadow, iv – Kobresia tundra, v – tundra-steppe, vi – cryophytic steppe, vii – petrophytic steppe, viii – dry steppe (bunch-grass typical and desert steppe).

We carried out floristic classification of plant communities of all basic types: dataset of 858 geobotanical releves (published information and unpublished materials of the authors) was analyzed. Most of releves belongs to published syntaxa, some new ones were ascertained on the basis of our unpublished data. The biodiversity of arid region vegetation is 25 associations and communities from 5 classes, 6 orders and 8 alliances (Table 1).

**Table 1.** Prodromus of basic plant communities of Altai-Sayan arid regions

Class **LOISELEURIO-VACCINIETEA** Eggler 1952
Order **BETULETALIA ROTUNDIFOLIAE** Mirkin et al. ex Chytrý, Pešout et Anenchonov 1993
Alliance ***Empetro-Betulion rotundifoliae*** Zhitlukhina et Onishchenko ex Chytrý, Pešout et Anenchonov 1993
1 Ass. ***Flavocetrario nivali-Betuletum rotundifoliae*** Telyatnikov 2013
2 Ass. ***Hylocomio splendentis-Caricetum sabynensis*** Telyatnikov 2013
3 Ass. ***Aulacomnio turgidi-Lagotidetum integrifoliae*** Telyatnikov 2013

Class **RHYTIDIO RUGOSI-LARICETEA SIBIRICAE** Korotkov et Ermakov 1999
Order **FESTUCO OVINAE-LARICETALIA SIBIRICAE** Korotkov et Ermakov ex Ermakov et al. 2000
Alliance ***Pachypleuro alpini-Laricion sibiricae*** Ermakov in Ermakov et al. 2000
4 Ass. ***Swertio obtusae-Laricetum sibiricae*** Makunina 2014
(syn. ***Polemonio coerulei-Laricetum sibiricae*** ass prov. Telyatnikov 2013)
5 Ass. ***Kobresio myosuroides-Laricetum sibiricae*** Ermakov in Ermakov et al. 2000
6 Ass. ***Artemisio rupestris-Laricetum sibiricae*** Makunina 2014

Class **JUNCETEA TRIFIDI** Hadac in Klika et Hadac 1944
Order **VIOLO ALTAICAE-FESTUCETALIA KRYLOVIANAE** ord. prov. Ermakov et Zibzeev 2012
Alliance ***Violo altaicae-Festucion krylovianae*** all. prov. Ermakov et Zibzeev 2012
7 Ass. ***Antennario dioicae-Pulsatilletum patentis*** Telyatnikov 2013
8 Ass. ***Seselo condensati-Trollietum asiatici*** Telyatnikov 2013

Class **CARICI RUPESTRIS-KOBRESIETEA BELLARDII** Ohba 1974
Order **KOBRESIETALIA MYOSUROIDIS** Mirkin et al. (1983) 1986
Alliance ***Kobresion myosuroidis*** Mirkin et al. (1983) 1986
9 Ass. ***Minuartio vernae-Papaveretum pseudocanescentis*** Telyatnikov et Mamakhatova 2011 (syn. ***Rhytidio rugosi-Oxytropidetum alpinae*** Telyatnikov 2013)
10 Ass. ***Festuco krylovii- Kobresietum myosuroidis*** ass. prov.
Alliance ***Potentillo niveae–Caricion pediformis*** Telyatnikov et Mamakhatova 2011
11 Ass. ***Eritrichio villosi-Helictotrichetum hookeri*** Telyatnikov et Mamakhatova 2011
12 Ass. ***Oxytropido recognitae–Helictotrichetum altaici*** ass. prov.
13 Ass. ***Carici rupestris–Festucetum krylovianae*** Korolyuk et Namzalov 1994

Class **CLEISTOGENETEA SQUARROSAE** Mirkin et al.1992
Order **HELICTOTRICHETALIA SHELLIANI** Hilbig 2000
Alliance ***Stellario petraeae–Festucion tschuensis*** all. prov.
14 Ass. ***Artemisio phaeolepidis-Kobresietum myosuroidis*** Makunina 2011
15 Ass. ***Sileno repentis–Caricetum pediformis*** Makunina 2014
16Ass. ***Stellario petraeae–Poetum attenuatae*** ass. prov.
17 Ass. ***Oxytropido eriocarpae-Poetum attenuatae*** Makunina 2014
18 Ass. ***Oxytropido macrosemae-Agropyretum cristatae*** Makunina 2014
Alliance ***Helictotrichion schelliani*** Hilbig 2000
19 Ass. ***Potentillo sericeae–Agropyretum cristati*** Makunina 2014
20 Comm. ***Oxytropis tragacantoides*** comm. prov.
Order **STIPETALIA KRYLOVII** Mirkin in Gogoleva et al. 1987
Alliance ***Stipion krylovii*** Kononov, Gogoleva et Mironova 1985
21 Ass. ***Artemisio frigidae-Stipetum krylovii*** Korolyuk et Makunina 2009
22 Ass. ***Artemisietum obtusilobae-santolinifoliae*** Korolyuk in Korolyuk et Makunina 2009
23 Comm. ***Stipa krylovii-Caragana bungei*** comm. prov.
24 Ass. ***Festuco tschuensis-Stipetum glareosae*** ass. prov. Korolyuk 2002
25 Ass. ***Allio vodopjanovae-Stipetum glareosae*** ass. prov. Korolyuk 2002

To find out the significance of differences of floristic classification syntaxa and their accordance with basic types of plant communities, we carried out a cluster analysis of floristic classification syntaxa. The dendrogram was made in Past [4] using Ward method.

Two main groups were revealed (Figure 2): 1) high mountain communities and cryophytic forest (i – v), 2) steppes (vi – viii). Shrub tundra (i) belongs to class **LOISELEURIO-VACCINIETEA**, cryophytic forest (ii) – to class **RHYTIDIO-LARICETEA.** Kobresia tundra (iv) and tundra-steppe (v) represent the class **CARICI-KOBRESIETEA**: the first one – alliance ***Kobresion myosuroidis***, the second one – alliance ***Potentillo niveae–Caricion pediformis.*** Alpine meadow (iii), because of the lack of regional higher syntaxonomic units, was pre-assigned to the class **JUNCETEA TRIFIDI.** Kobresia tundra, tundra-steppe and alpine meadow form one common cluster. All steppe communities were ascribed to class **CLEISTOGENETEA SQUARROSAE**. Dry steppe was included in the order **STIPETALIA KRYLOVII**, the others – in the order **HELICTOTRICHETALIA SHELLIANI:** petrophytic steppe belongs to alliance ***Helictotrichion schelliani***, cryophytic one – to alliance ***Stellario petraeae–Festucion tschuensis*** .

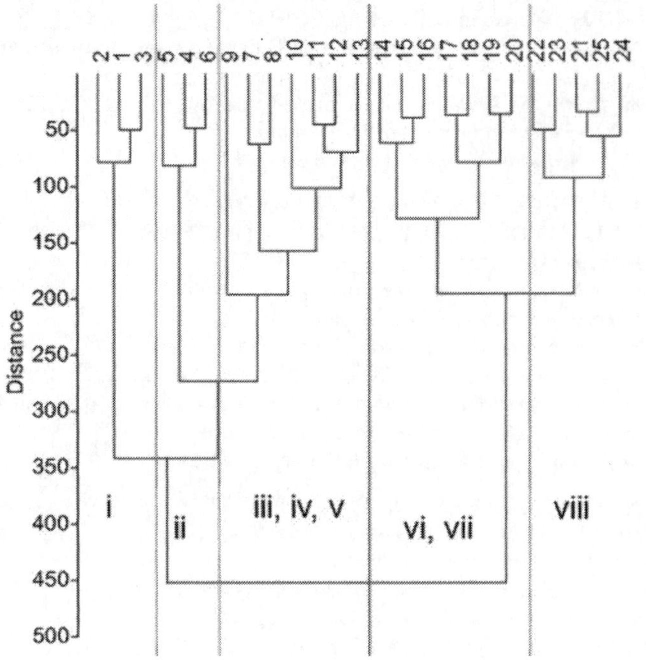


**Fig. 2.** Dendrogram of syntaxa. Arabic numerals in the dendrogram correspond to those in the prodromus. i – shrub tundra, ii – cryophytic forests, iii – alpine meadows, 4 – Kobresia tundra, 5 – tundra-steppes, 6 – cryophytic steppes, 7 – petrophytic steppes, 8 – bunch-grass steppes.

## 1.2 Characteristics of basic plant communities

To give short characteristics of basic plant communities we chosen active species, defining community visual appearance. The activity of the species was calculated as the square root of the multiplication of constancy by mean cover [5]. Very active (activity ≥40), active (40> activity≥ 20), and moderately active (20 > activity ≥ 10) species form the core of active species (Table 2). Very active species dominate in plant communities, active ones – codominate; moderately active species are constantly present.

*Betula rotundifolia*, mosses (*Aulacomnium turgidum, Hylocomium splendens, Rhytidium rugosum*), and lichens (*Cladonia arbuscula, C. stellaris, C. amaurocrea)* define the appearance of shrub tundra (i). *Larix sibirica* dominates in tree layer of cryophytic forest (ii); *Betula rotundifolia* and *Lonicera altaica* form shrub layer; tolerant to cold species make up the herb layer. Alpine meadow (iii) has no obvious dominants; *Pentaphylloides frticosa*, a wide range of alpine species, and tolerant to cold meadow-steppe species (*Pulsatilla patens*) codominate. Kobresia tundra (iv) has two main dominants: *Kobresia myosuroides* and *Dryas oxyodontha*. Tundra-steppe (v) and cryophytic (vi) steppe occur only in arid high mountains; they possess the numerous group of cryophytic species, the most part of which are inactive ones. Their distinctive feature is the simultaneous presence of species of different altitudinal zone groups. Alpine (*Carex rupestris, Festuca kryloviana*), forest-steppe (*Carex pediformis*), and steppe (*Helictotrichon altaicum*) species codominate in tundra-steppes (v); cryophytic species *Eremogone meyeri* and *Saussurea schanginiana* are moderately active. Steppe (*Koeleria cristata, Helictotrichon altaicum*) and forest-steppe (*Carex pediformis*) species make the core of cryophytic steppe (vi), alpine species have no significant role; in numerous group of cryophytic species only *Festuca tschujensis* is active. In petrophytic steppe (vii) steppe grasses (*Agropyron cristatum, Poa botrioides*) dominate; numerous petrophytic herbs are present, but not active. Dry steppe (viii) are composed of only steppe species (*Agropyron cristatum, Artemisia frigida, Poa botryoides*).

**Table 2.** Active species of basic plants communities in Altai-Sayan arid regions

| Basic vegetation types | i | ii | iii | iv | v | vi | vii | viii |
|---|---|---|---|---|---|---|---|---|
| Most of plant communities | | | | | | | | |
| *Aster alpinus* | | x | xx | x | x | x | x | . |
| *Carex pediformis* | . | x | x | x | xx | xx | x | |
| *Helictotrichon altaicum* | | . | x | . | xx | xx | xx | |
| *Poa botryoides+P. attenuata* | . | x | x | x | xx | xx | xx | x |
| Highmountain and cryophytic forest communities | | | | | | | | |
| *Betula rotundifolia* | xx x | xx | x | x | | | | |
| *Bistorta major* | x | . | xx | xx | x | . | | |
| *Bistorta vivipara* | x | x | x | xx | . | . | | |
| *Carex stenocarpa* | x | . | x | xx | . | . | | |
| *Cetraria islandica* | xx | . | x | x | . | . | | |
| *Cladonia arbuscula* | xx | . | . | x | | | | |
| *Festuca altaica* | x | x | xx | x | . | . | | |
| *Flavocetraria cucullata* | x | . | . | xx | . | . | | |
| *Hedysarum austrosibiricum* | x | x | xx | x | . | | | |
| *Rhytidium rugosum* | xx | xx | . | xx | . | . | | |
| *Spiraea alpina* | x | . | x | x | . | . | | |
| Shrub tundra | | | | | | | | |
| *Aulacomnium turgidum* | xx | . | | . | | | | |
| *Cladonia stellaris* | xx | | | . | | | | |
| *Dicranum species* | xx | . | | . | | | | |
| *Hylocomium splendens* | xx | . | | | | | | |
| *Cladonia amaurocraea* | x | | | . | | | | |
| *Cladonia rangiferina* | x | | | . | | | | |
| *Peltigera species* | x | | . | . | | | | |
| *Polytrichum strictum* | x | . | | | | | | |
| *Pyrola incarnata* | x | . | . | . | | | | |
| *Salix glauca* | x | . | . | . | | | | |
| *Sanionia uncinata* | x | | | | | | | |
| Cryophytic forests | | | | | | | | |
| *Larix sibirica* | . | xxx | . | . | | . | | |
| *Festuca ovina+ F.* | x | xx | | | | | | |
| *Abietinella abietina* | | x | | . | | . | | |
| *Aconitum baicalense* | . | x | | | | | | |
| *Aconitum barbatum* | | x | . | | . | . | | |
| *Atragene sibirica* | . | x | . | | . | . | | |
| *Bromopsis pumpelliana* | | x | . | . | . | . | | |
| *Carex amgunensis* | | x | | | | . | | |
| *Lonicera altaica* | . | x | . | | | . | | |
| *Polemonium* | . | x | . | . | | . | | |

| Basic vegetation types | i | ii | Iii | iv | v | vi | vii | viii |
|---|---|---|---|---|---|---|---|---|
| Cryophytic forests, alpine meadows and Kobresia tundra | | | | | | | | |
| *Bupleurum multinerve* | | x | X | x | . | . | . | |
| *Kobresia myosuroides* | . | x | X | xxx | . | . | | |
| *Pachypleurum alpinum* | . | x | X | x | . | . | | |
| *Potentilla nivea* | | x | X | x | . | | | |
| Alpine meadows, Kobresia tundra and tundra-steppes | | | | | | | | |
| *Carex rupestris* | . | | X | x | xx | | | |
| *Festuca kryloviana* | . | | Xx | xx | xx | | | |
| Alpine meadows and Kobresia tundra | | | | | | | | |
| *Dracocephalum grandiflorum* | . | . | Xx | x | . | . | | |
| *Eritrichium villosum* | . | . | X | x | . | . | | |
| *Helictotrichon hookeri* | | . | X | x | . | . | . | |
| *Minuartia verna* | . | . | X | x | . | . | . | . |
| *Potentilla gelida* | . | . | X | x | . | . | | |
| *Schulzia crinita* | . | . | X | x | . | . | | |
| *Thalictrum alpinum* | . | . | X | x | . | . | | |
| Kobresia tundra | | | | | | | | |
| *Dryas oxyodonta* | . | . | . | xxx | . | . | | |
| *Flavocetraria nivalis* | . | | . | xx | . | . | . | |
| *Oxytropis alpina* | . | | . | xx | . | | | |
| *Carex ledebouriana* | . | . | . | x | . | . | | |
| *Claytonia joanneana* | . | | . | x | | . | | |
| *Crepis chrysantha* | . | . | . | x | . | | | |
| *Gentiana algida* | . | | . | x | . | | | |
| *Gentiana grandiflora* | . | . | . | x | . | . | | |
| *Lagotis integrifolia* | . | . | . | x | . | . | | |
| *Papaver pseudocanescens* | . | . | . | x | . | . | | |
| *Patrinia sibirica* | | . | . | x | . | . | . | |
| *Pedicularis oederi* | . | . | . | x | . | . | | |
| *Ptilagrostis mongholica* | | . | . | x | . | . | , | |
| *Thamnolia vermicularis* | . | | | x | . | | | |
| Kobresia tundra and tundra-steppe | | | | | | | | |
| *Eremogone meyeri* | | . | . | x | x | | | |
| *Saussurea schanginiana* | | . | . | x | x | . | | |
| Tundra-steppes | | | | | | | | |
| *Agrostis tuvinica* | | . | . | . | x | . | | |
| *Artemisia phaeolepis* | | . | . | . | x | . | . | . |
| Steppes | | | | | | | | |
| *Artemisia frigida* | | . | . | | . | x | xx | xx |
| *Koeleria cristata* | | . | . | . | . | xx | x | x |
| Cryophytic and petrophytic steppes | | | | | | | | |

| | i | ii | iii | iv | v | vi | vii | viii | | i | ii | iii | iv | v | vi | vii | viii |
|---|---|---|---|---|---|---|---|---|---|---|---|---|---|---|---|---|---|
| *pulcherrimum* | | | | | | | | | *Potentilla sericea* | | . | . | . | . | x | x | . |
| *Saussurea alpina* | . | x | . | . | . | . | | | Cryophytic steppes | | | | | | | | |
| Cryophytic forests and alpine meadows | | | | | | | | | *Festuca tschujensis* | | | | | | xx | | . |
| *Cotoneaster uniflorus* | | x | x | . | . | . | . | | *Artemisia pycnorhiza* | | | . | . | . | x | . | |
| *Dianthus superbus* | . | x | x | . | . | . | | | *Astragalus multicaulis* | . | . | . | . | . | x | . | . |
| *Geranium pseudosibiricum* | | x | x | . | | | | | *Pulsatilla ambigua* | | | . | . | . | x | . | . |
| *Myosotis imitata* | . | x | x | . | . | . | | | Petrophytic steppes | | | | | | | | |
| *Poa sibirica* | . | x | x | . | . | | | | *Festuca valesiaca* | | | | | | | xx | |
| *Tephroseris praticola* | . | x | x | . | . | . | | | *Artemisia santolinifolia* | | | . | | . | . | x | . |
| Alpine meadows | | | | | | | | | *Oxytropis tragacanthoides* | | | | | | . | x | . |
| *Pentaphylloides fruticosa* | . | . | xx | . | . | . | | . | Dry and petrophytic steppes | | | | | | | | |
| *Pulsatilla patens* | . | . | x | . | . | . | | | *Agropyron cristatum* | | | | | | . | xx | xx |
| *Aconogonon diffusum* | . | . | x | . | . | . | . | . | *Potentilla acaulis* | | | | | . | . | x | x |
| *Antennaria dioica* | . | . | x | . | . | . | | | Dry steppes | | | | | | | | |
| *Artemisia rupestris* | | . | x | . | . | . | . | | *Caragana bungei* | | | | | | | . | x |
| *Erigeron eriocalyx* | | . | x | . | . | . | | | *Carex duriuscula* | | | | | | . | . | x |
| *Gentiana uniflora* | | . | x | . | | . | | | *Kochia prostrata* | | | | | | . | . | x |
| *Ligularia altaica* | | . | x | | . | . | . | | *Stipa glareosa* | | | | | | | | x |
| *Seseli condensatum* | | . | x | . | . | . | | | *Stipa krylovii* | | | | | | . | . | x |
| *Swertia obtusa* | . | . | x | . | | | | | | | | | | | | | |
| *Trollius asiaticus* | . | . | x | . | | | | | | | | | | | | | |

**Notes**. xxx – very active species, xx – active species, x – moderately active species, "." – inactive species. Basic types of plant communities: i – shrub tundra, ii – cryophytic forest, iii – alpine meadow, iv – Kobresia tundra, v – tundra-steppe, vi – cryophytic steppe, vii – petrophytic steppe, viii – bunch-grass steppe.

### 1.3 Vegetation structure

The steppe belt (1800–2400 m) is confined to the Chuya and Kargy depressions and the adjacent flanks of mountain ranges. The lower subbelt (1800–2100m) occupies the bottoms of depressions. The background plant community is dry steppe (viii). There are two main forms of relief here. The first form occurs on mountain flanks of the Chuya depression. Steep rocky slopes and screes dominate; the background vegetation is petrophytic steppe (vii). The second form of relief predominates in arid regions; the background plant community is cryophytic larch forest (ii) and cryophitic steppe (vi). In the arid conditions of regions under discussion, forests can exist only due to the moisture received in summer during the melting of the ice permafrost. Cryophytic larch forest (ii) covers concave slopes, while cryophytic steppe (vi) occupies convex ones. In the lower part of the alpine belt (2400–2600 m), the slopes of shadow exposures are covered with shrub tundra (i); the sunny slopes – with Kobresia tundra (iv) and tundra-steppe (v). Tundra-steppe is confined to the stony part of the slopes; Kobresia tundra occupies areas with more developed soils. In the vegetation cover of the upper part of alpine belt (2600–2800 m), Kobresia tundra absolutely dominates. Alpine meadows (iii) occur throughout all alpine belt.

## 3 Conclusion

Thus, 8 basic types of plant communities were revealed in Altai-Sayan arid region vegetation. Shrub tundra was included in class **LOISELEURIO-VACCINIETEA**, cryophytic forest – in class **RHYTIDIO-LARICETEA**; alpine meadow was pre-assigned to class **JUNCETEA TRIFIDI.** Kobresia tundra and tundra-steppe represent class

**CARICI-KOBRESIETEA**; steppes were ascribed to class **CLEISTOGENETEA SQUARROSAE**.

The authors are grateful to the Russian Foundation for Basic Research for providing financial support to the research (grant № 18-04-00822 A).

## References

1. I.S. Novikov, *Morphotectonics of the Altai* (Publishing House of Siberian Branch of RAS, Novosibirsk, 2004)
2. V.V. Sevastyanov, *Climatic resources of Altai mountains and their applications* (Publishing House of Tomsk Pedagogical University, Tomsk, 2009)
3. N.I. Makunina, The fourth international scientific conference "Ecology and geography of plants and plant communities" (2012)
4. Ø. Hammer, *PAST Paleontological Statistics. Version 2.17. Reference manual* (University of Oslo, Oslo, 2012)
5. L.I. Malyshev, Botanical J., **58** (1973)

# The «red snow» of the Polar Urals (Western Siberia)

*Yuriy* Naumenko*

Central Siberian Botanical Garden, SB RAS, 630090, Zolotodolinskaja Str., 101, Novosibirsk, Russia

**Abstract**: Pink-red colored snow fields were sampled in the area of Ochety Lake (the Polar Urals, West Siberia) at the altitude of 272 m above the sea level in August 2019. Zygospores of *Chlamydomonas nivalis* prevailed in plant communities. Altogether, 9 species of algae have been discovered in snow samples: 7 species of Cyanoprokaryota, 1 species of Bacillariophyta and 1 species of Chlorophyta.

Tinted snow is observed rather frequently in the mountains and polar regions. Red snow was first scientifically mentioned by Horace de Saussure, a Swiss natural scientist, in 1760. He discovered it at the foot of the Alpes in the Duchy of Savoy in France [1]. Analogous findings were later noted in different areas of the Alpes, Pyrenees, in the north of Scandinavia, in Alyaska, in polar countries and other places. I. V. Palibin was the first who mention pink and red snow in the European part of Russia. In 1901, he took part in the polar expedition on "Yermak" icebreaker, where he studied flora of Franz Josef Land, Spitzbergen and Severny (Northern) island of Novaya Zemlya. He observed the red snow phenomenon on Novaya Zemlya near Mashigin Bay and in the area of Matochkin Strait where pink snow is encountered in ravines, mountain slopes, almost everywhere.

Research demonstrated that the red tint of snow is caused by the development of *Chlamydomonas nivalis* (F.A.Bauer) Wille algae. Communities of this algae was encountered in high mountain snow fields in polar regions all over the world. Snow algae are cold-resistant organisms that manage to grow on snow and ice. It is known from literature that algae caused "snow blooming". Dependently on the prevalence of certain types of snow algae and other organisms the tint can be different. Powerfully vegetating algae tints snow fields and ice into different colours: orange, yellow, yellowish-green, green, blue, brown [2].

Snow algae live in an environment which is called cryobiotop. Cryobionts are certain group of algae which vegetate in snow and ice in appropriate conditions. The ecology of snow algae is highly important for understanding snow and ice ecosystems. However, there is not enough environmental information about snow algae.

In 1928, T.K. Tripolitova [3] described the presence of *Chlamydomonas nivalis* (F.A.Bauer) Wille in West Siberia. She noted that the specie is frequently found in the nival zone of Altay mountain ridges, but did not provide any references to the

* Corresponding author: Naumenko55@ngs.ru

researchers. Later N.N. Voronikhin provided data about the presence of this specie in the mountains of Polar Urals [4], based on plant collecting of 1925-1927.

## Materials and methods

Snow samples werecollected in the first decade of August 2018 around Ochety Lake. The lake is located in Priuralsky district of Yamalo-Nenets Autonomous Okrug. This water body is located in a narrow intermontane valley. Multiple residuals of snow cover (Fig. A-B) were found on the mountain slopes and in the ravines during the research period. Water surface area is 1.8 km$^2$, it is a flow-through lake, a lot of creeks fall into it. The areas of the tinted snows varied from 0.5 to 20 m$^2$, the samples were collected from the surface with the depth of not more than 5–7 cm. Geographical coordinates of sampling plot were 68°04'02"N, 65°50'26" E; 272 m. above the sea level. The material was analized under Amplival optical microscope with magnification 16 x 40. Samples were collected by I.A. Gorbunova (Central Siberian Botanical Garden). The author expresses deep appreciation for her contribution.

**Fig**. **A** - Ochety Lake; **B** - tinted snow

## Results and discussion

Red-snow algae prevailed in the snow samples and it caused the tinting of snow into pink and red. The specie Chlamydomonas nivalis (F.A.Bauer) Wille (Syn. *Uredo nivalis* F.A. Bauer, *Sphaerella nivalis* (F.A. Bauer) Sommerfelt, *Palmella nivalis* (F.A.Bauer) W.J.Hooker *Protococcus nivalis* (F.A.Bauer) Agardh, *Coccochloris nivalis* (F.A.Bauer) Sprengel, *Coccophysium nivale* (F.A.Bauer) Link, *Haematococcus nivalis* (F.A.Bauer) Morren) is part of the family Chlamydomonadaceae of Chlorophyta taxonomic section [5]. Cells are ellipsoidal or widely ooid, from posterior and anterior view (less frequently only from a posterior view) they are widely round. The membrane is thick, layered, is uniformly thickened along the whole cell. The red pigment astaxanthin usually masks chlorophyll.

While studying the samples we discovered spherical cells with thick cellular wall surrounded by semi-transparent mucus (aplanospores) sometimes covered with mineral particles. Analogous phenomenon was observed by D. Remias et al. [6]. The cell content is red. Cilia have not been discovered in the studied material. This was a dominat specie in cryosecton of the Polar Urals. Besides, round cells (zygospores) of this specie with stellate external membranes and small round-shaped cells have been discovered. The latter were encountered much more frequently. The sizes of the cells with stellate external membranes (see the Table.) varied from 20.0 to 30.0 mym (the average diameter is 21.9 mym). The diameter of small round cells without ornamentation changed from 12.5 to 17.5 mym (the average dimeter is 15.4 mym). N. N. Voronikhin stated in his research (1930) that he discovered cells with uniform diameter 9.0-10.8 mym in the sample collected at the head of the river Kozhima.

**Table.** The linear sizes of Chlamydomonas nivalis cells collected in August 2018

| Characteristic feature of a cell | n, unit | Actual limits, mym | Mo, mym | $\bar{x} \pm s_{\bar{x}}$, mym | S mym | υ, % |
|---|---|---|---|---|---|---|
| Cells with ornamentation | 31 | 20.0–30.0 | 20 | 21.86±0.57 | 3.16 | 14 |
| Small cells | 35 | 12.5–17.5 | 15 | 15.36±0.29 | 1.73 | 11 |

**Note**. N – number of variants in a given sample; $\bar{x}$ – average value of a characteristic; Mo – modal value of a characteristic; S – standart deviation of sampled population; $s_{\bar{x}}$ – statistical error; υ – coefficient of variation.

Besides the dominat specie, 7 species of Cyanoprokaryotes were discovered in snow samples: *Aphanocapsa muscicola* (Menegh.) Wille (=Microcystis muscicola (Menegh.) Elenk.), *Gloeocapsa alpina* Näg., *Chroococcus minutus* (Kütz.) Näg. (=Gloeocapsa minuta (Kütz.) Hollerb.), *C. turgidus* (Kütz.) Näg. (=*Gloeocapsa turgida* (Kütz.) Hollerb.), *Gloeocapsopsis magma* (Brйb.) Kom. et Anagn. (=*Gloeocapsa magma* (Breb.) Kütz.), *Leptolyngbya foveolarum* (Rabenh. ex Gom.) Anagn. et Kom. (=Phormidium foveolarum Gom.), *L. frigida* (Fritsch) Anagn. et Kom. (=*Phormidium frigidum* Fritsch, *Pseudanabaena frigida* (Fritsch) Anagn.). The Bacillariophytes were represented by one specie: *Hantzschia amphyoxis* (Ehr.) Grun. All abovementioned species were encountered singly. It is worth noting that the discovered species, according to the literature, are widely represented both in hydrophyte and soil communities. At the same time, a few species relate to a greater degree to cryobionts in terms of their ecology. Thus J. Kvíderová [7] noted *Leptolyngbya foveolarum* while she was studying the snow fields of Central Spitsbergen. In 1930, G. S. Filippova [8] found *Chroococcus minutus* and *Hantzschia amphyoxis* in the snow fields of the Caucasus, the latter has also been discovered in the snow fields in the mountains of Bulgaria [9]. The study of criobiontic organisms is highly important not only for botany but also for biotechnologists. These organisms contain valuable compounds such as astaxanthin and its esters, and unsaturated fatty acids. Astaxanthin protects from ultraviolet rays and is the scavenger of free radicals [6].

## Conclusion

Thus, the color of the snow in the Polar Urals is caused by intense growth of *Chlamydomonas nivalis*. In the studied snow samples from Ochety Lake 9 species of algae have been discovered, including 7 species of the Cyanoprokaryotes, 1 species of the Bacillariophytes and 1 species of the Chlorophytes.

## References

1. A.G. Salimovskaya–Rodina, Archives of biological Sciences, **43**, 2-3 (1936).
2. B. Kawecka, Polish Polar Research, **7**, 4 (1986).
3. T.K. Tripolitova, Izv. Tomsk state University, **79**, 4 (1928).
4. N.N. Voronikhin, Proceedings of the Leningrad region of natural scientists, **60**, 3 (1930).

5. M.D. Guiry, G.M. Guiry, AlgaeBase, World-wide electronic publication, National University of Ireland, Galway. (2019) http://www.algaebase.org; searched on 31 July 2019.
6. D. Remias, U. Lütz–Meindl, C. Lütz, Eur. J. Phycol, **40** (2005).
7. J. Kvíderová, Czech Polar Reports, **2** (1) (2012).
8. G.S. Filippova, Izv. USSR Academy of Sciences, **7**, 7 (1934)
9. J. Lukavský, V. Cepák, Nova Hedwigia, **99**, 3–4 (2014).

BIO Web of Conferences **16**, 00023 (2019) https://doi.org/10.1051/bioconf/20191600023
*Results and Prospects of Geobotanical Research in Siberia*

# Standard flora as a standard quantitative evaluation of vascular plant diversity of terrestrial ecosystems

*Evgenii* G. Nikolin[1, 2*]

[1]Institute for Biological Problems of Cryolithozone Siberian Branch of RAS, Lenina ave., 41, Yakutsk, 677980, Russia
[2]State Nature Reserve Ust-Lensky, Acad. Fedorova, 28, 678400, Tiksi, Bulunsky ulus, Republic of Sakha (Yakutia), Russia

**Abstract**. A new method of accounting for the diversity of vascular plants in the reference area of standard flora, covering an area of 1 sq. km, has been tested in the basins of the Lena and Indigirka rivers. Quantitative indices of the diversity of 3 standard floras in different latitudinal and altitudinal ranges of North-East Asia are presented.

Quantitative assessment of vascular plant diversity of a particular region is usually based on the method of specific flora studying developed by A. I. Tolmachev [1, 2] and his followers [3-5]. A.I. Tolmachev initially proposed the area of 100 km square for a specific flora. However, the allocation of this area depends on a variety of geographical and other factors, and is determined subjectively by the researcher. As a result, the boundaries of specific or local flora become vague, and quantitative indicators of the diversity of different floristic areas are comparable with a high degree of conditionality. It is very difficult to investigate a complete flora in the area of 100 sq. km during one summer season. Besides, no one researcher can assure that he keeps the boundaries of a selected transect strictly. Owing to this, additionally to commonly approved methods of a specific flora study within its natural boundaries (which are admeasured conventionally), I offer to emphasize on a 1 sq. km plot as a standard flora (SF) unit for thorough investigation of vascular plants diversity [6, 7]. Before going to the research area, it is advisable to pre-schedule the site under the standard flora and set the coordinates of its boundaries. For example, you can use a public Google Earth resource to do this. On the sites, you comply the boundaries of this transect uses an electronic Navigator (GPS). For the study of standard flora it is preferable to choose a smooth, viewable areas (river valleys, saddles, passes, plateau). On the borders of the standard flora, on the ground it is desirable to establish noticeable landmarks. The selected area should be investigated as fully as possible. It is convenient when the standard flora is on the border or surrounds the base camp. In the lists of specific flora of the surveyed area, the species within the standard flora should be marked with an apostrophe or other sign. It is assumed that the investigated area can serve as a standard for monitoring observations of changes in the composition of the flora.

* Corresponding author: enikolin@yandex.ru

At the moment, this method has been tested, some material has been accumulated, data on 2 sites in the Lena river basin and 1 site in the Indigirka river basin have been published. The first site is located in the middle reaches of the Lena river, within the subzone of the middle taiga forests [8, 9], on the Kharyyalakh Island (SF coordinates: 61°04'21.34" N and 126°20'46.97" E – 61°04'53.50" N and 126°20'28.22" E – 61°05'01.03" N and 126°21'33.44" E – 61°04'28.49" N and 126°21'49.68" E) [7]. The second section is located in the Lena River Delta, on the right Bank of the Bykovskaya duct, it covers the territory adjacent to the International Biological Station "Lena-Nordenskjold", located within the subzone of the Arctic tundra (SF coordinates: 72°11'48.97" N and 128°01'34.14" E – 72°11'50.72" N and 128°03'21.59" E – 72°11'18.95" N and 128°03'14.17" E – 72°11'16.21" N 128°01'32.36" E) [10]. The third site is located in the headwaters of the Artyk-Yuryue stream (the Indigirka river basin), in the Tas-Kistabit range (SF coordinates: 64°27'26.6" N and 143°21'35.2" E – 64°27'36.5" N and 143°22'49.2" E – 64°27'53.1" N and 143°20'49.9" E – 64°28'02.0" N and 143°22'03.2" E) [11]. In addition to the latitudinal position, biodiversity is also affected by the altitude factor. The surveyed areas, according to the electronic resource Google Earth, are located in the following high-altitude limits: 1 – 110 – 115; 2 – 10 – 25; 3 – 1100 – 1300 m above sea level. The specific flora of Kharyyalakh island is stated to comprise 299 species of vascular plants, while its SF includes 228 species. The specific flora of the Lena River Delta is represented by 258 species of vascular plants, and its SF by 170 taxons (161 species, 1 hybrid and 8 subspecies). In the North-East of Yakutia, in the range of Tas-Kistabit, in the specific flora of Nelkanskogo mountain there are 122 total identified taxons (114 species of, 1 hybrid, 3 subspecies and 4 varieties), of which SF is noted 92 taxons (87 species, 3 subspecies and 2 varieties).

It is clear that these data are indicative of an overall decline in diversity to the North and with elevation. It is too early to analyze them in depth, it is necessary to accumulate such data for a more objective assessment of biodiversity of different territories. It would be desirable that such observations were supported by botanists from other regions. This would help to have more objective assess to the diversity of vascular plants in different geographical and environmental conditions and to understand the patterns of its formation.

The work was carried out according to the draft State assignment of the Institute for Biological Problems of Cryolithozone of the Siberian Branch of the Russian Academy of Sciences No. AAAA-A17-117020110056-0.

## References

1. A.I. Tolmachev, Journal RBO. **16**(1) (1931)
2. A.I. Tolmachev, *Methods of comparative floristics and problems of florogenesis* (Novosibirsk: Nauka, 1986)
3. B.A. Yurcev, Botanical journal, **60**, 1 (1975)
4. Yu.R. Shelyag-Sosonko, Botanical journal, **65**, 6 (1980)
5. R.V. Kamelin, Comparative floristic: species diversity analysis of plants. Problems. Perspective. "Tolmachevsky readings" (2014)
6. E.G. Nikolin, Comparative floristic: species diversity analysis of plants. Problems. Perspective. "Tolmachevsky readings" (2014)
7. E.G. Nikolin, *Concrete, partial and standard flora of Central and North-Eastern Yakutia* (Novosibirsk: Nauka, 2015)
8. *The main features of the vegetation cover of the Yakut ASSR* (Yakutsk: YaF SO AN SSSR, 1987)

9. E.I. Troeva, A.P. Isaev, M.M. Cherosov, N.S. Karpov, Plant and vegetation, **3** (2010)
10. E.G. Nikolin, I.A. Yakshina, V.V. Petrovsky, Botanical journal, **102**, 10 (2017)
1. E.G. Nikolin, Botanical journal, **104**, 3 (2019)

# Ecological features and new location of the rare relict species *Craniospermum pseudotuvinicum* (Boraginaceae)

*Svetlana* Ovchinnikova*

Central Siberian Botanical Garden, SB RAS, 630090, Novosibirsk, Zolotodolinskaya str. 101, Russia

**Abstract.** Information on the new location of the high-mountainous species *Craniospermum pseudotuvinicum* (endemic to Mongolia) is given. The morphological and ecological-cenotic features of the species, as well as known locations within the Central Mongolian-Altai province are considered.

When performing a critical revision of taxa of the family Boraginaceae Juss. on the territory of Outer Mongolia and the study of the extensive herbarium collections of the Central Asia Department of the Herbarium of the Botanical Institute V. L. Komarova of the Russian Academy of Sciences (LE), we found a specimen (not previously defined), which we assigned to the recently described species Craniospermum pseudotuvinicum Ovczinnikova et A. Korolyuk: "Outer Mongolia, the Great Lakes Basin, m. Batar Khairkhan (eastern part), alt. 2500 m above sea level, southern exposition, on scree, sedge-petrophyte-forb community, 26 VII 1983, N 255, E.A. Volkova, I.Yu. Sumerina, U. Beckett, H. Buyan-Orshy"(LE).

The found plants are perennial stem-root with many-headed caudex loose frutescentes monopodially growing polycarpic with a main vegetative rosette shoot and elongated axillary generative shoots. Rosette leaves are back-lanceolate, rounded at the top, 8–12 cm long., 1.0–2.0 cm wide, densely pubescent with thin long, 2–2.5 mm, half-adherent gray hairs and short soft hairs. Flowering stems are 10 cm tall, simple, distant hairy, leafy to the inflorescence. Inflorescence is a double coil, almost capitate, 3.3 cm long. Calyx is mildly tangled-white-haired with back-lanceolate lobes, 7.0–8.5 mm long. Corolla is lilac, saffron when dry, 9.0–10.0 mm long, dolichomorphic. Corolla teeth are 1.3 mm long. Anthers are 1.1–1.2 mm long., slightly exposed from the corolla. The stigma is hidden inside the corolla. The plant is flowering, the fruits are not known.

The analysis showed that the characteristics of the life form, the pubescence of the leaves and calyx, the signs of the inflorescence and the flower, the size of the anthers, and the ecological features of the plants correspond to the characteristics of C. pseudotuvinicum.

Species of the genus Craniospermum Lehm. are perennial dense-custy grasses, common in semi-desert-steppe belt of Central Kazakhstan, southern Siberia, Mongolia, and Northern China, from 42° to 55° N. As shown by palynomorphological studies, the genus

* Corresponding author: sv-ovchin@yandex.ru

Craniospermum can be attributed to the earliest representatives of the ancient Mediterranean flora [1]. Currently, the genus includes 8 species that are part of 3 sections: C. subvillosum Lehm. (section Craniospermum) is an endemic of the littoral of Lake Baikal within the Irkutsk Region and the Republic of Buryatia; C. canescens DC. and C. desertorum Ovczinnikova et A. Korolyuk (section Leucolachnos Ovczinnikova) are endemics of the Russian and Mongolian Altai; C. subfloccosum Krylov (subsection Floccosa Ovczinnikova of the section Diploloma (Schrenk) Popov) is a narrow endemic of the Russian (Katunsky Ridge) and the Kazakhstan Altai (Narymsky Ridge); C. echioides (Schrenk) Bunge (section Diploloma) is an endemic of Dzungaria within the three states of Kazakhstan, Mongolia and China; C. mongolicum I.M. Johnston (section Diploloma) is endemic to Mongolia and China within northern Xinjiang and Inner Mongolia; C. tuvinicum Ovczinnikova (section Diploloma) is endemic to Tyva and Mongolian Altai; C. pseudotuvinicum Ovczinnikova et A. Korolyuk (section Diploloma) is a narrow endemic of the Mongolian Altai. All species of the genus Craniospermum are found on dry gravelly slopes and scree, in the crevices of rocks in the belt of high-altitude and upland-xerophyte steppes, in their ecological confinement cryoxerophytes [2-5].

The species C. pseudotuvinicum was described from the eastern foothills of the Zhargalant-Khairkhan-Ula massif within the Chandman somon of the Khobdos aimag (holotype - NSK0000710). The classical location and newly discovered one are located in the Great Lakes Basin region (Fig. 1). Locations are indicated on a map made with SimpleMappr (http://www.simplemappr.net).

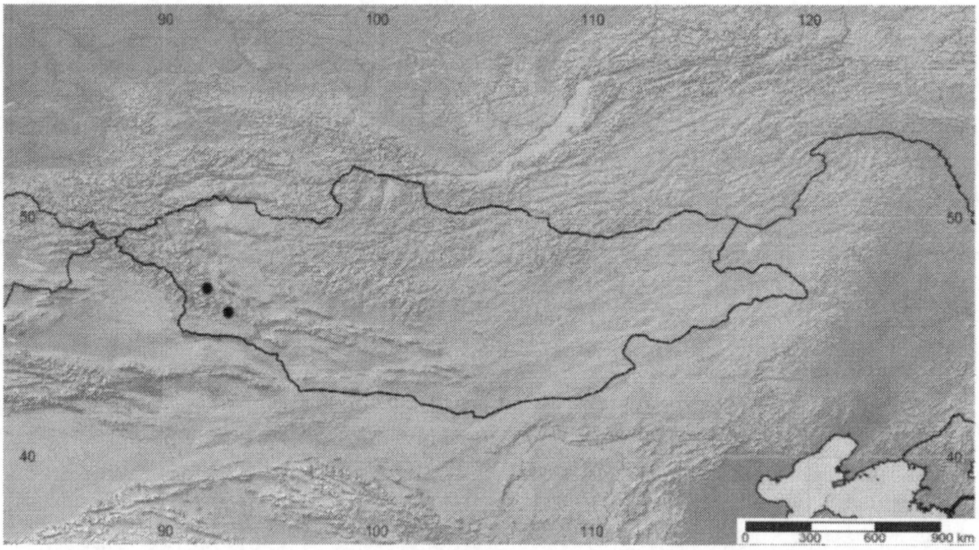


**Fig. 1.** Map of the areal of *Craniospermum pseudotuvinicum*.

The eastern part of the Batar-Khairkhan Ridge and the eastern foot of the town of Zhargalant-Khairkhan-Ula are located at an altitude of 2400–2550 m above sea level within the Central Mongolian-Altai province in the region dominated by dry sod-grassland steppes, the usual dominants of which are widespread steppe species: Poa attenuata Trin., Koeleria cristata (L.) Pers, Stipa krylovii Roshev. Habitats are mainly sloping with poorly developed stony soils with high activity of petrophytic species: Festuca lenensis Drobov, Eremogone meyerii (Fenzl) Ikonn., Amblynotus rupestris (Pall. Ex Georgi) Popov ex Seg., Orostachys spinosa (L.) C.A. Mey. At heights of 2600 m and more the grass-sedge cryophytic steppes begin to play an important role in vegetation cover. These communities are characterized by high abundance of Festuca kryloviana Reverd., Poa attenuata Trin., Carex rupestris All., Oxytropis oligantha Bunge, Potentilla nivea L. and Eremogone meyerii (Fenzl) Ikonn. Due to the considerable dissection of the relief, unclosed groups of petrophytic plants are widespread at all heights, among which Craniospermum pseudotuvinicum.

The main types of communities of the Central Mongolian-Altai province are characteristic of Mongolia or Central Asia as a whole, but a rather large number of communities are endemic Mongol-Altai or even Central-Mongol-Altai types [5, 6].

The work was performed as part of the state assignment of the Central Siberian Botanical Garden of the Siberian Branch of the Russian Academy of Sciences on the project AAAA-A17-117012610055-3 "Biological diversity of cryptogamic organisms (algae, fungi, lichens) and vascular plants in the geospace of biotic and abiotic factors, their role in aquatic and terrestrial ecosystems of North Asia". In preparing the publication, materials of the bioresource scientific collection of CSBG SB RAS "Herbarium of Higher Vascular Plants, Lichens and Fungi (NS, NSK)", UNU № USU 440537 were used.

## References

1. S.V. Ovchinnikova, Bot. Journ., **86**, 12 (2001)
2. S.V. Ovchinnikova, Bot. Journ., **85**, 12 (2000)
3. S.V. Ovchinnikova, Turczaninowia, **22** (to be published)
4. S.V. Ovchinnikova, V.P. Orlov, Bot. Journ., **90**, 9 (2005)
5. S.V. Ovchinnikova, A.Yu. Korolyuk, Plant Life of Asian Russia, **3**(23) (2016)
6. E.A. Volkova, *Botanical geography of Mongolian and Gobian Altai* (Komarov Botanical Institute, St. Petersburg, 1994)

# The Impact of Climate Change on *Betula tortuosa* Ledeb. Radial Increment on the Eastern Macroslope of Kuznetsk Alatau

*Il'ya* A. Petrov[1,*], *Alexey* S. Golyukov [1,2], *Alexandr* S. Shushpanov [1,3], and *Viacheslav* I. Kharuk [1,2]

[1]V. N. Sukachev Institute of Forest SB RAS, 660036, Krasnoyarsk, Akademgorodok, 50/28, Russia
[2]Siberian Federal University, 660041, Krasnoyarsk, Svobodny pr., 79, Russia
[3]Reshetnev Siberian State University of Science and Technology, 660037, Krasnoyarsk, Krasnoyarsky Rabochy Av., 31, Russia

**Abstract.** The climatic response of birch in eastern Kuznetzky Alatau Mountains (Siberia) were analysed based on dendrochronology data. Studied climate variables included temperature, precipitation and root zone moisture (RZM). Periods with maximal correlations for each variable were found. It was established that radial increment was limited by June temperature and July RZM. Correlations of radial increment and eco-climatic factors were higher when a negative anomalies of respective factor occurred. To estimate the advancement rate of birch trees along elevation gradient, three elevational transects were laid in mountain forest-tundra ecotone. Estimated advancement rate is about 0.5 m per year. Although temperature have a stimulating effect on birch's radial increment and advancement on elevation gradient in Kuznetzky Alatau Mountains, some depressions in radial increment corresponds to years with soil moisture decrease.

## 1 Introduction

Climate change has a multidirectional effect on forest ecosystems. At treeline, the observed temperature increase mainly has a stimulating effect on the radial increment and the advancement of trees along the elevation gradient [1, 2, 3]. At the same time, aridization of the climate at lower elevations, caused by an increase in temperature, leads to a massive decline of dark coniferous forests [4]. Under such conditions, ecotones (transition zones between different types of vegetation) are the most sensitive areas to climatic impacts. Trees growing in ecotones especially strongly react to changes in the temperature and moisture regime [5]. The reaction of birch (*Betula tortuosa* Ledeb.) trees to climatic changes in the mountain forest-tundra ecotone of the Kuznetsk Alatau Mountains is also of great interest because in the Altai-Sayan region communities with birch are unique objects [6]. Our work aimed to analyse the effects of recent climate change on the birch (*Betula tortuosa* Ledeb.) trees.

[*] Corresponding author: petrovilsoran@gmail.com

## 2 Materials and methods

Studied birch trees grew on eastern macroslope of Kuznetsk Alatau Mountains at elevations from 1100 to 1500 m above sea level (fig. 1). Samples for dendrochronological analysis (N = 111) were collected at four main sites. To estimate the rate of advancement, three transects were laid from timberline to treeline. On each transect temporary test plots (3X3 or 5X5 m) with an interval of 10 m along the elevation gradient were laid to describe morphometric parameters of trees, some soil and vegetation features. Sampled trees were selected throughout each transect. For each sampled tree, coordinates and elevation were fixed.

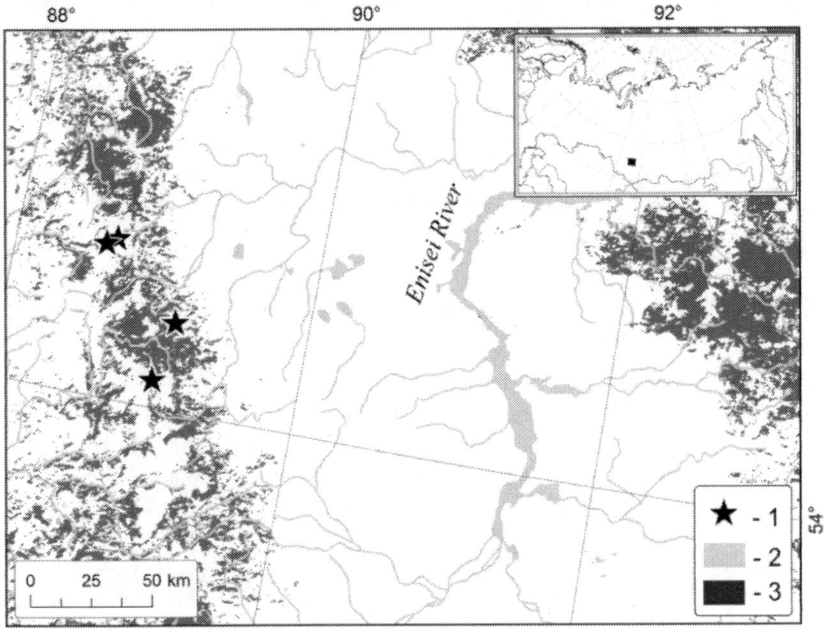


**Fig. 1.** Geographical location of the studied sites (1 – test plots, 2 – rivers and lakes, 3 – dark needle conifer stands).

Dendrochronological analysis was performed according to generally accepted methods [7]. Measurement of tree-ring width was carried out on the platform LINTAB-6. The quality of cross-dating was tested in the COFECHA program [8]. Tree-ring chronologies were indexed in the ARSTAN program using the negative exponent or linear regression with a negative slope methods [9]. Radial increment in final chronology is presented in dimensionless growth indexes (GI) with an average of 1.0 and a relatively constant variance. The average interserial correlation coefficient of the chronologies is 0.61; average sensitivity coefficient is 0.44. The series of eco-climatic parameters were obtained from the "Nenastnaya" weather station (WMO 29752; about 10-60 km from test sites) and the database of the MERRA-2 project. Pearson's correlation coefficients (including running 11-yr correlation) were used in dendroclimatic analysis.

## 3 Results

Tree-ring chronology of *Betula tortuosa* trees have an increasing ($r^2$ = 0.78) trend in the period 1968-1977 and a negative trend ($r^2$ = 0.64) in 1994–2000 (fig. 2). Similar dynamics in radial increment was recorded earlier in *Larix sibirica* Ledeb. trees growing in the study area [3]. Between the chronologies of these species, a significant correlation coefficient was found (0.55; period 1940-2012).

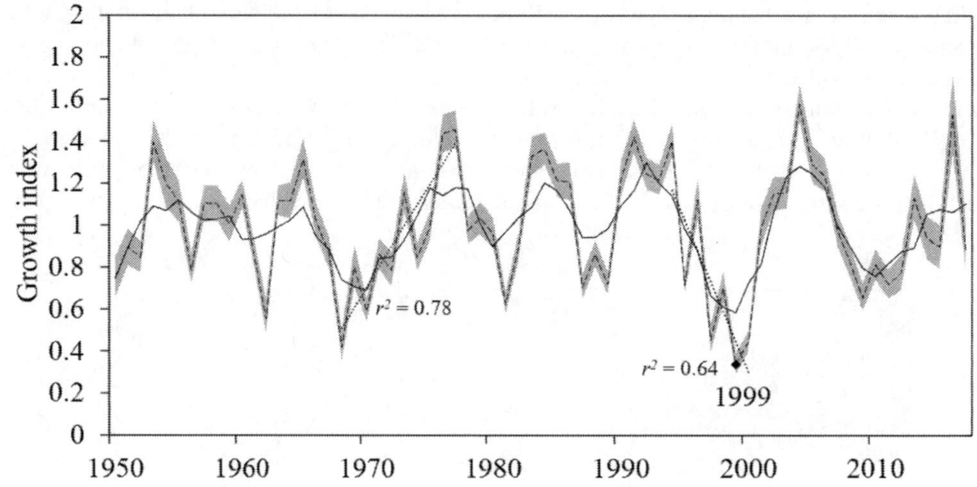


**Fig. 2.** Tree-ring chronology of *Betula tortuosa* trees (dot indicated absolute minimum of radial increment; grey background indicated confidence level ($p < 0.05$); solid line indicated 5-yr running average).

The radial increment of birch correlates with the temperature of June, however, the climatic signal is unstable (fig. 3a), achieved in some periods r = 0.8. Significant correlations coincide with periods of June temperature negative anomalies. For example, the decrease in June temperature explains the fluctuations of the radial increment in the period from the early 1980s to the early 1990s (June temperature was 1.4 °C lower than average). However, correlation between radial increment and June temperatures is insignificant in the period of a GI decrease in 1990s (with minimum in 1999).

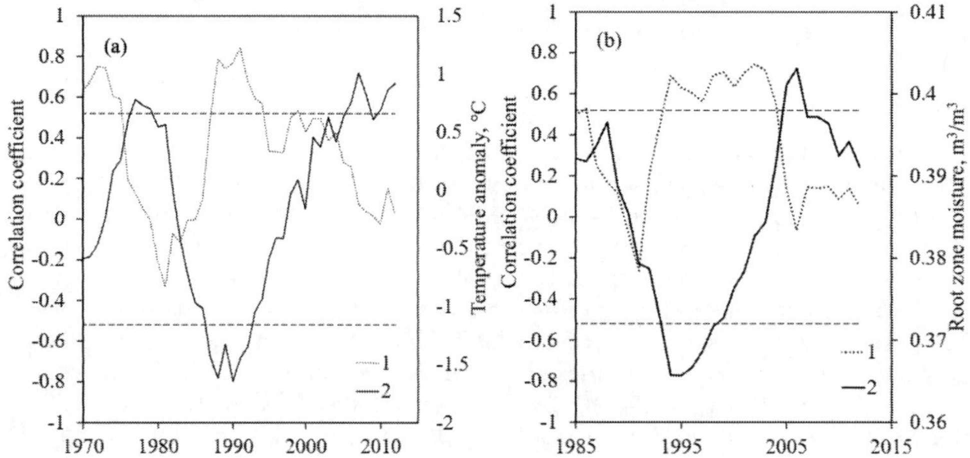


**Fig. 3.** Dynamics of 11-yr running correlation between GI and eco-climatic parameters (1) and 11-yr running average of eco-climatic parameters (2): a – June temperature anomaly; b – July root zone moisture (dashed line - $p < 0.05$).

Period about 1995 characterized by low root zone moisture (RZM; fig. 3b). GI decrease in late 1990s corresponds to soil drought (in 1999 both GI and RZM reached the minimum). Correlations of July RZM and GI become stable significant during soil moisture decrease (r = 0.65; fig. 3b).

Based on data of tree's age and coordinates, rate of birch advancement estimated about 0.25-0.5 m per year. Field data confirm that birch trees have appeared at elevations not previously occupied by woody plants.

During the period of climatic changes (1970-2018), radial increment of *Betula tortuosa*, growing in the Kuznetsk Alatau Mountains, is mainly correlated with air temperature in June and root zone moisture in July. Correlations increase in a period of lowering the parameters of corresponding eco-climatic factor. Temperature increase also stimulating advancement of birch trees along elevation gradient.

## 4 Findings

- The rate of *Betula tortuosa* advancement along the elevation gradient is estimated ~ 0.25-0.5 m / year.
- Radial increment of *Betula tortuosa* trees is influenced by June temperature and July root zone moisture.
- The magnitude of *Betula tortuosa* climatic response increases in a period of lowering the parameters of the corresponding eco-climatic factor.

Acknowledgements. The Russian Science Foundation (17–74–10113) supported dendrochronological and dendroclimatic analysis. The Russian Foundation for Basic Research supported a part of analysis of ecological and climatic variables (grant 18–45–240003).

## References

1. M. Harsch, P. Hulme, M. McGlone, R. Duncan, Ecol. Lett., **12 (1)** (2009)
2. V.I. Kharuk, S.T. Im, M.L. Dvinskaya et al., J. Mountain Sci., **14(3)** (2017)
3. I.A. Petrov, V.I. Kharuk, M. L. Dvinskaya, S.T. Im, Contemp. Probl. Ecol. **8(4)** (2015)
4. V.I. Kharuk, S.T. Im, I.A. Petrov et al., Reg. Environ. Change, **17** (2017)
5. F.K. Holtmeier, *Mountain Timberlines: Ecology, Patchiness, and Dynamics*, (Dordrecht, Netherlands, Kluwer, 2009).
6. E.G. Zibzeev, E.A. Basargin, T.A. Nedovesova, Veg. Rus., **26** (2015)
7. J.H. Speer, *Fundamentals of tree-ring research* (Tucson, Univ. of Arizona Press, 2010)
8. R.L. Holmes, Tree-ring Bull., **44** (1983)
9. E.R. Cook, R.L. Holmes, *Guide for computer program ARSTAN* (Lab. of Tree Ring Research, University of Arizona, 1986)

BIO Web of Conferences **16**, 00026 (2019) https://doi.org/10.1051/bioconf/20191600026
*Results and Prospects of Geobotanical Research in Siberia*

# Mosses in *Brachypodio-Betuletea* forests of West Siberia

*Olga* Pisarenko*

Central Siberian Botanical Garden SB RAS, 630090, Zolotodolinskaya Str., 101, Novosibirsk, Russia

**Abstract.** *Brachypodio-Betuletea* forests of West Siberia were evaluated for the mosses. The observed localities are in limits 51°48' – 56°48' N and 60°16' – 88°20' E. 93 moss species are registered their, but only 36 are common. Most of the species are associated with *Betula*-trunks. The list of species and their occurrence by main types of the forests are given. The diversity and abundance of mosses correlate with the community humidity, decreasing with xerophytization. It is shown that on the studied territory moss species composition and constancy well indicate the latitudinal replacement of classes of forest vegetation.

The southern periphery of the forest zone in West Siberia, toward south of 57-56° N, is composed of birch and aspen-birch grass forests. The forests are regarded to be indigenous here. They are considered as vicariant to broad-leaved and coniferous-broad-leaved forests of the ocean regions [1, 2]. In the Braun-Blanquet approach, they are allocated in the class ***Brachypodio pinnati-Betuletea pendulae*** Ermakov et al. 1991. Forests with the dominance or participation of pine on the territory are azonal and restricted by sandy or gravelly substrates and, accordingly, take place mainly on the ancient high terraces of large rivers and in ancient runoff ravines. The pine forests are also included into class ***Brachypodio-Betuletea.***

Vast gradient of heat and moisture values from North to South and regional variation in flora cause significant difference in the composition of the grass and shrub layers in the discussed forests. As result a lot of associations of floristic classification were described. Meanwhile, today for the territory there is no prodromus of vegetation unites.

Bryophyte investigation of the forests was carried out in the 2009-2016 years, together with geobotanical ones; 252 plots were examined on the territory of Novosibirsk, Kemerovo, Omsk, Tyumen provinces and Altai Territory in limits 51°48' – 56°48' N and 60°16' – 88°20' E.

In sum in the investigated spots 93 moss species were recorded; 67 of them were found more then once (table), but only 36 are more or less common.

Meticulous bryophyte investigations of Bashkiria forests were shown the absence diagnostic mosses for particular forest associations, the specificity of bryopvhyte composition of individual associations is extremely low [3]. Therefore, to describe the

* Corresponding author: o_pisarenko@mail.ru

distribution of mosses, I integrated ecologically and geographically similar associaton and arranged the examined forest communities into 6 groups according their habitat specifity, species composition and distribution. For the listed below groups I indicated their syntaxonomic position; abbreviation used in the table is given in brackets.

– moist tall-herbaceous birch and birch-aspen forests; common in sub-taiga; order *Carici macrourae-Pinetalia sylvestris,* alliance *Lathyro gmelinii-Pinion sylvestris* (b1).
– mesophytic birch and aspen-birch herbaceous forests; common in sub-taiga and forest-steppe of the south-eastern part of West Siberia and foothills of Altai-Sayan region; order *Calamagrostio epigeii-Betuletalia pendulae,* alliance *Calamagrostio epigeii-Betulion pendulae* (b2).
– mesophytic and moderately dry birch herbaceous forests of Baraba forest-steppe zone; order *Carici macrourae-Pinetalia sylvestris,* alliance *Vicio unijugae-Pinion sylvestris* (b3).
– moderately dry birch herbaceous forests of forest-steppe zone West of the Ob-River; order *Calamagrostio epigeii-Betuletalia pendulae,* alliance *Peucedano morisonii-Betulion pendulae* (b4).
– mesophytic pine and birch-pine shrub-herbaceous forests of Ob-River terraces and ancient runoff ravines; order *Calamagrostio epigeii-Betuletalia pendulae,* alliance *Peucedano morisonii-Betulion pendulae* (p1).
– mesophytic pine and birch-pine shrub-herbaceous forests of Salair ridge; order *Carici macrourae-Pinetalia sylvestris,* alliance *Vicio unijugae-Pinion sylvestris* (p2).

All investigated types of birch and aspen forests are similar in set of constant moss species (table, upper part). The differences mainly related to rare and random species (table, lower part). The diversity and abundance of mosses correlate with the community humidity and decrease with xerophytization. So in the most moist *Lathyro gmelinii-Pinion* forests 64 moss species were recorded; in *Vicio unijugae-Pinion* forests – 61 species, in *Calamagrostio epigeii-Betulion* forests – 42; in *Peucedano morisonii-Betulion* – 16. In all types of herbaceous small-leaved forests the main ecotope for mosses are trunks of the dominant tree, *Betula pendula* or *B. pubescens*. Soil surface in the forests under close herb layer is covered by hay and forest litter; a dead-fall usually is dry and not populated by mosses or completely absent due to periodic ground fires. Moss abundance on tree trunks depends on many factors: the thickness and inclination of the trunk, total humidity, snow depth in winter, position of the community/tree in the relief. In old-growth birch forests of sub-taiga zone on trunks about 40 cm diameter mosses grow up to 60-80 cm above ground. In dry insular birch groves of forest-steppe zone the range is reduced to 15-20 cm.

A set of mosses living in pine forests is almost the same as in small-leaved ones (table). Only a few species grow on soil: *Pleurozium schreberi* is the most common; *Ptilium crista-castrensis* and *Dicranum polysetum* take place; other 'taiga species' – *Hylocomium splendens Rhytidiadelphus triquetrus* and *Polytrichum commune* – are relatively rare. In birch and aspen forests all moss species sometimes occur on tree trunks. Moss diversity in pine forests is caused by species growing on tree trunks. Most species were found on birches which are usually solitary present in the pine forests. In the result moss species lists of small-leaved and pine forests of *Brachipodio-Betuletea* on the territory are very similar (table).

In forest zone of West Siberia *Betula pendula* and *B. pubescens* are widespread from its southern to northern limits. In subtaiga and forest-steppe zones the presence of *Betula*-trees provides the similarity of moss lists in different types of forest communities. But further south and north bryophyte component of forest communities undergoes appreciable changes. Geographic ranges of many epiphytic species extends further north to Middle and North taiga zones and even to Forest-Tundra. But afore, in the Southern taiga zone, most of them (*Platygyrium repens, Callicladium haldanianum, Stereodon pallescens, Dicranum montanum,* etc.) disappear or sharply reduce the constancy and abundance in zonal communities and occur mainly in valley forests [4]. Even more drastic changes in bryophyte population of forest communities occur in the southern direction. Toward south limit of forest-steppe subzone, small-leaved forests lose their zonal positions; they occur on spots with higher humidity. Species composition of these forests differs from this one of the communities of *Brachypodio-Betuletea* [2]. These forests are extremely poor in mosses. Only 13 species were recorded there. They have contrast ecological features. Five of them are never noted in the discussed *Brachypodio-Betuletea* communities (*Bryum creberrimum, Leptodictyum riparium, Pterygoneurum kozlovii, Tortula muralis, Warnstorfia fluitans*).

So, in south of West Siberia moss species composition and constancy well indicate the latitudinal replacement of classes of forest vegetation.

**Table**. Occurrence of moss species in investigated types of *Brachypodio-Betuletea* forests.

| | Community type | | | | | |
|---|---|---|---|---|---|---|
| | **p1** | **p2** | **b1** | **b2** | **b3** | **b4** |
| **Number of releves** | 55 | 20 | 67 | 21 | 102 | 6 |
| **Number of moss species** | 67 | 41 | 64 | 42 | 61 | 16 |
| *Pylaisia polyantha* | III | III | V | V | V | V |
| *Amblystegium serpens* | IV | + | IV | V | V | V |
| *Ceratodon purpureus* | II | I | I | II | II | III |
| *Stereodon pallescens* | III | III | III | III | IV | V |
| *Orthotrichum speciosum* | II | + | III | II | III | III |
| *Sanionia uncinata* | IV | III | IV | III | IV | IV |
| *Brachythecium salebrosum* | IV | IV | IV | IV | III | I |
| *Platygyrium repens* | II | + | III | III | III | I |
| *Brachythecium capillaceum* | II | . | III | III | III | II |
| *Sciuro-hypnum reflexum* | III | I | III | + | III | II |
| *Callicladium haldanianum* | IV | II | III | III | III | II |
| *Dicranum montanum* | II | III | III | III | II | II |
| *Pohlia nutans* | III | II | II | III | II | IV |
| *Orthotrichum obtusifolium* | I | + | III | III | II | II |
| *Plagiothecium denticulatum* | I | + | I | + | II | II |
| *Pylaisia selwynii* | . | . | II | I | I | I |
| *Plagiomnium cuspidatum* | IV | II | IV | III | IV | . |
| *Pleurozium schreberi* | III | V | II | III | II | . |
| *Haplocladium microphyllum* | II | . | II | III | II | . |
| *Dicranum fuscescens* | + | IV | II | II | II | . |
| *Sciuro-hypnum curtum* | III | + | II | + | I | . |
| *Dicranum polysetum* | II | IV | II | II | + | . |
| *Ptilium crista-castrensis* | II | V | II | I | + | . |
| *Rhytidiadelphus triquetrus* | I | + | II | + | + | . |
| *Campylidium sommerfeltii* | I | + | I | + | + | . |
| *Plagiomnium ellipticum* | + | . | + | + | + | . |
| *Oncophorus wahlenbergii* | + | + | I | I | + | . |
| *Dicranum flagellare* | + | + | + | I | + | . |
| *Climacium dendroides* | + | II | + | + | + | . |
| *Polytrichum juniperinum* | I | I | + | + | + | . |
| *Calliergonella lindbergii* | . | . | + | + | + | . |
| *Plagiothecium laetum* | + | + | + | + | + | . |
| *Leskea polycarpa* | II | . | I | . | II | . |
| *Plagiomnium medium* | + | II | + | + | . | . |
| *Rhodobryum roseum* | + | II | + | . | + | . |

| | | | | | | |
|---|---|---|---|---|---|---|
| *Funaria hygrometrica* | I | . | . | + | + | . |
| *Leptobryum pyriforme* | I | I | . | . | + | . |
| *Dicranum scoparium* | + | . | . | I | + | . |
| *Abietinella abietina* | + | . | + | . | + | . |
| *Atrichum flavisetum* | I | + | . | . | + | . |
| *Aulacomnium palustre* | + | . | + | . | . | . |
| *Barbula unguiculata* | + | . | . | + | + | . |
| *Brachythecium baicalense* | + | + | . | . | . | . |
| *Brachythecium mildeanum* | + | . | + | + | . | . |
| *Brachythecium rutabulum* | + | . | + | . | . | . |
| *Bryoerythrophyllum recurvirostrum* | + | + | . | . | + | . |
| *Bryum caespiticium* | + | + | . | . | + | . |
| *Campyliadelphus chrysophyllus* | + | . | . | . | + | . |
| *Dicranum bonjeanii* | . | . | + | . | + | . |
| *Eurhynchiastrum pulchellum* | + | . | + | . | + | . |
| *Fissidens bryoides* | + | . | + | . | . | . |
| *Homalia trichomanoides* | + | . | + | . | + | . |
| *Hygroamblystegium varium* | + | . | . | + | + | . |
| *Hylocomium splendens* | + | + | + | . | . | . |
| *Mnium stellare* | + | . | + | . | . | . |
| *Oxyrrhynchium hians* | + | . | + | . | . | . |
| *Pohlia cruda* | + | . | + | . | + | . |
| *Polytrichum commune* | . | + | . | . | + | . |
| *Polytrichum piliferum* | + | + | . | . | + | . |
| *Pseudoleskeella nervosa* | . | . | + | . | + | . |
| *Rhytidiadelphus subpinnatus* | . | . | + | + | . | . |
| *Sciuro-hypnum starkei* | + | + | + | + | . | . |
| *Serpoleskea subtilis* | . | . | + | + | . | . |
| *Tetraphis pellucida* | . | + | + | . | . | . |
| *Thuidium recognitum* | + | . | + | + | . | . |
| *Timmia megapolitana* | + | . | + | . | + | . |
| *Weissia brachycarpa* | + | . | + | . | . | . |

**Note.** Species with solitary records are not included. Nomenclature follows to [5]. Community types are explained in the text. Occurrence classes (ranges of the percentage of releves in which the species is registered): V – 81-100%; IV – 61-80%, III – 41-60%; II – 21-40%, I – 11-20%, + less than 10%.

The work response to state assignment AAAA-A17-117012610052-2 and biocollection development № USU 440537, partly supported by RFFI №18-04-00822.

## References

1. N.N. Lavrenko, *The vegetation of the West Siberian plain* (Novosibirsk, 1985)
2. N.N. Lashchinskiy, The vegetation of East Europe and North Asia, **87** (2014)
3. E.Z. Baisheva, V.B. Martynenko, P.S. Shirokikh, *Bryophyta of forest ecosystems of Bashkortostan Republic* (Ufa, Gilem, Bashk. Enzikl, 2015)
4. O.Yu. Pisarenko, Plant world of Asian Russia, **2(14),** 24-28 (2014)
5. M.S. Ignatov, O.M. Afonina, E.A. Ignatova et al., Arctoa, **15** (2006)

# The lichens of the plant communities in the middle reaches of the Uty River Valley

*Alena* Sagalakova, *Olga* Zyryanova*, *Mariya* Larina

Khakas State University named after N.F. Katanov, 90 Lenin St., 655000 Abakan, Russia

**Abstract.** The article presents the data on the species communities situated in the middle reaches of the Uty river valley (Beisky District, Republic of Khakassia). The taxonomic list of lichens is represented by 51 taxa of a species rank belonging to 24 genera and 11 families. The largest number of species is characterized by the crustose life form. Mesophytes and epiphytes are predominant in the terms of ecological substrate.

## 1. Introduction

Lichens are a completely inimitable and extremely peculiar group of living organisms. They always attract attention, primarily because they grow in the places where other plants cannot grow and, therefore, they are called "pioneers" of vegetation. At the same time they remain a poorly studied group of plants [1]. Studying lichens is especially relevant as lichens are the most sensitive components of ecosystem due to the fact that even a moderate atmospheric pollution adversely affects them causing a certain reduction in species diversity and foliage projective cover up to the complete extinction of many species [2].

The Uty river is located in the Beisky District and forms the right tributary of the Abakan river. The valley vegetation is represented by bunchgrass, meadow and salt steppes, valley meadows, windbreaks and forest outliers.

According to the available literature there is no information on the studies of the species composition of lichens in the Uty river valley. In this regard, the species composition of lichens in the plant communities situated in the middle reaches of the Uty river valley has particular interest for further studies.

## 2. Material and methods

The materials for the study include the herbarium specimen of lichens collected by the authors in the middle reaches of the Uty river valley with the use of route method. The research was conducted on the southern slope of Mount Bolshaya, on the territory of the village Uty, in the floodplain of the Uty river, in the wood lines and on the different substrata. The material collection and herbarization was performed using the common method described in the second edition of Identification Guide to Lichens of the USSR "Morphology, Systematics and Geographical Distribution" [3]. Around 60 herbarium envelopes were collected and then processed in the Laboratory of the Khakas State

* Corresponding author: o_a_zyryanova@mail.ru

University named after N.F. Katanov with the use of the microscope Mikromed-1 and binocular loupe MBS-2 with 10 % KOH solution. In order to identify the materials "Identification Guide to Lichens of the USSR" [4-7] and "Identification Guide to Lichens of Russia" [8-12] were used. The life forms and ecological groups were identified on the basis of the personal observations of the authors and according to the literary sources [13].

## 3. Results and discussion

The species diversity of lichens in the middle reaches of the Uty river valley is represented by 11 families, 24 genera, including 51 species: *Cyphelium tigillare* (Ach.) Ach., *Candelaria concolor* (Dick.) Stein., *Candelariella aurella* (Hoffm.) Zahlbr., *C. vitellina* (Hoffm.) Müll. Arg., *C. xanthostigma* (Ach.) Lettau., *Cladonia pocillum* (Ach.) Grognot., *Cl. pyxidata* (L.) Hoffm., *Aspicilia cinerea* (L) Körb. *A. maculata* (H. Magn.) Oxner., *A. transbaicalia* Oxner., *Lecanora bolcana* Pollini., *L. campestris* (Schaer.) Hue., *L. cenisia* Ach., *L. crenulata* Hook., *L. dispersa* (Pers.) Sommerf., *L. frustulosa* (Dicks.) Ach., *L. hageni* Ach., *L. subfusca* Ach., *Lecidella euphorea* (Flörke) Hertel., *Rhizoplaca chrysoleuca* (Sm.) Zopf., *Rh. melanophtalma* (DC.) Leuckert et Poelt. Rh. peltata (Ramond) Leuckert et Poelt., *Flavoparmelia caperata* (L.) Hale., *Flavopunctelia soredica* (Nyl.) Hale, *Melanelia tomini* (Oxner) Essl., *Parmelia cetrata* Müll. Arg., *P. olivacea* (L.) Essl., *P. sulcata* Taylor., *Parmeliopsis ambigua* (Wulfen) Nyl., *Photoparmeliopsis muralis* (Schreb.) M.Choisy., *Xanthoparmelia camschadalis* (Ach.) Hale., *X. conspersa* (Ach.) Hale., *X. somloënsis* (Gyeln.) Hale., *Pertusaria pertusa* (Weigel) Tuck., *Buellia insignis* (Naegeli ex Hepp) Körb., *Diplotomma alboatrum* (Hoffm.) Flot., *Phaeophyscia sguarrosa* Hale., *Physcia aipolia* (Ehrh. ex Humb.) Fürnr., *Ph. albinea (Ach.) Nyl., Ph. caesia* (Hoffm.) Fürnr., *Ph. leptalea (Ach.) DC., Physcia obscura (Aht.)* Rabenh., *Ph. stellaris* (L.) Nyl., *Ph. tenella* (Scop.) DC., *Rinodina septentrionalis* Malme., *Psora decipiens* (Hedw.) Hoffm., *Rhizocarpon disporum* (Nageli ex Hepp.) Müll. Arg., *Caloplaca ferruginea* (Huds.) Th. Fr., *C. flavorubescens* (Huds.) J. R. Laundon., *C. pellodella* (Nyl.) Hasse., *Xanthoria elegans* (Link) Th.Fr.

The average number of species in a family is approximately equal to 4.6, while 3 families have values above this indicator and considered dominant. They include Lecanoraceae *Körb.* (12; 23.5 %), Physciaceae Zahlbr. and Parmeliaceae Zenker (11 species; 21.5 %). Candelariaceae Hakul. и Teloschistaceae Zahlbr include 4 species (7.8 % each). There are also 3 single-species families on the studied territory which make 5.8 % of the total species number.

The high position in the flora composition of Lecanoraceae Körb., Teloschistaceae Zahlbr. families is characteristic of the arid regions flora in the Holarctic [14]. At the same time a significant content of Parmeliaceae Zenker. and Physciaceae Zahlbr families makes it similar to the boreal floras of the Holarctic.

The detailed layout of lichenoflora can be inferred by the spectrum of the most-represented genera in the terms of species.

The average number of species in genus is 2.17. There are 7 dominant genera exceeding this value and accounting for 68.6 % of the total species number. The largest number of species belongs to the genus of Lecanora (Ach.) Th. Fr. with 8 species which make 15.6 %. The representatives of these species typically include the steppe species, as, for example, *Lecanora crenulata* and *L. frustulosa*. The next genus in terms of number is Physcia (Schreb.) Michx. having 7 species which make 13.7 %. The typical representatives are *Physcia stellaris* and *Ph. tenella* located on the bodies of the deciduous trees, especially on birches.

The genera Aspicilia A. Massal., Rhizoplaca Zopf, Parmelia Ach., Caloplaca Th. Fr. and Xanthoparmelia (Vain.) Hale have 3 species which make 29.4 % of the total species number. The predominance of the mentioned genera indicates that the studied area is located in the steppe belt.

There are 13 (24.7 %) single-species genera within the territory of the Uty river valley.

The biomorphological analysis has shown that the most species of lichens have a crustose form of thallus. They are 26 in number which is 51% of the total species number.

This number includes such species as *Candelariella aurella*, *Protoparmeliopsis muralis*, *Lecanora crenulata* and others. The number of species having a foliose form of thallus within the studied area is 22, which is 43.1 %. This number includes such species as *Xanthoparmelia somloënsis*, *Physcia stellaris*, *Parmeliopsis ambigua*, *Xanthoparmelia conspersa* and others. The lichens having a fruticose form of thallus are represented by 3 species which makes 5.9 % from the total number of the collected species. *Aspicilia transbaicalica* have a dwarf-fruticose form, but nevertheless the authors classified it as fruticose type.

In view of the fact that crustose lichens hold the dominant position, according to the classification by A.N. Oksner [3], in addition to the three main types one can distinguish their biomorphs characterized as more fractional. They include areolate-crustose, granular-crustose and scaly-crustose lichens. The largest number of lichens among crustose type found in the middle reaches of the Uty river valley has the areolate (areolate-crustose) type of thallus and they are mainly confined to the stony substrate of the steppe and forest-steppe zones.

To characterize the ecological structure of lichens the authors used the data on the lichens tolerance to humidity, thermal regime, and snow cover thickness according to the work of N.V. Sedelnikova [13]. As a result, all collected lichens were divided into 3 ecological groups. The ecological group of mesophytes includes 29 species which is 57 % of the total number. These are all species of the genus Physcia, as well as species such as *Rhizocarpon disporum*, *Pertusaria pertusa*, *Lecidella euphorea* and others. The group of xerophytes includes 18 species which is 35.3 % and comprises such species as *Aspicilia maculata*, *Lecanora campestris*, *Lecanora subfusca* and others. The smallest number is represented by the group of xerocryophytes of 4 species which is 7.8 %. This group includes species from the genus Rhizoplaca and Melanelia tomini.

According to the substrate analysis it is established that lichens located within the studied area prefer a stony (26.5 %) and woody (21.4 %) types of substrate. Only 4 species grow on the soil which is 7.8 % of the total species number. The largest number of epilithic and epiphytic lichens is characteristic of the studied area since the steppe type of vegetation with wood lines, planted with poplars is considered predominant. Besides, a dominant group of epiphytic lichens can be found right there. There are also stony steppes serving a substrate for epilithic lichens.

## 5. Conclusion

The territory under study belongs to the steppe zone, where fruticose lichens are not large in number. This fact for its part confirms our analysis on the life forms of lichens.

The studies were supported by the RFBR projects 18-04-00633 and 18-44-190007 p_a.

## References

1. T.N. Pystina. *Botany problems at a boundary 20-21 cent.*, **2** (St. Petersburg, 1998)
2. A.V. Altynnikova, V.V. Reutskaya, *Actual problems of environmental protection and environmental management* (Stavropol, 2014)

3. A.N. Oksner. *Identification Guide to Lichens of the USSR,* **2** (Science, Leningrad, 1974)
4. *Identification Guide to Lichens of the USSR,* **1** (Science, Leningrad, 1971)
5. *Identification Guide to Lichens of the USSR*, **3** (Science, Leningrad, 1975)
6. *Identification Guide to Lichens of the USSR*, **4** (Science, Leningrad, 1977)
7. *Identification Guide to Lichens of the USSR*, **5** (Science, Leningrad, 1978)
8. *Identification Guide to Lichens of Russia*, **6** (Science, St. Petersburg, 1996)
9. *Identification Guide to Lichens of Russia*, **7** (Science, St. Petersburg, 1998)
10. *Identification Guide to Lichens of Russia*, **8** (Science, St. Petersburg, 2003)
11. *Identification Guide to Lichens of Russia*, **9** (Science, St. Petersburg, 2004)
12. *Identification Guide to Lichens of Russia*, **10** (Science, St. Petersburg, 2008)
13. N.V. Sedelnikova, *Lichens of Altai and Kuznetsk highlands: Summary of flora* (Science, Novosibirsk, 1990)
14. N.S. Golubkova, *Analysis of lichen flora of Mongolia* (L., Science, 1983)

BIO Web of Conferences **16**, 00028 (2019) https://doi.org/10.1051/bioconf/20191600028
*Results and Prospects of Geobotanical Research in Siberia*

# Space structure and diversity of island vegetation on Torey lakes in drought period in Daursky reserve

*Lyudmila* Saraeva*

Daursky State Nature Biosphere Reserve, Nizhniy Tsasuchey, 76 Komsomolskaya Str., Zabaikalsky krai, Russia

**Abstract.** The Torey lakes of Daursky reserve belong to wetlands of international importance and are located in the south of Zabaikalsky krai. In 2009, in the drought period, we studied the space structure and diversity of the vegetation on five islands in the Torey lakes. The results of ecological-phytocenotic classification are presented. 52 associations related to 15 formations and 4 types of island vegetation on the Torey lakes are identified.

## 1. Introduction

The Daursky nature reserve with the area of 49,764 ha, and the buffer zone of 173,201 ha is located in the south of Transbaikal territory at the border with Mongolia. The reserve includes the Torey lakes inscribed in the list of wetlands of international importance under Ramsar convention. The Torey lakes consist of soda lakes Barun-Torey and Zun-Torey. The Torey lakes are at the intersection of two powerful bird flyways - the Eastasian-Australasian and Eastasian-Indian ones. Barun-Torey Lake feeds the steppe Ulda and Imalka rivers. Zun-Torey Lake feeds atmospheric precipitation and the groundwater to the surface. In the aquatorium of Torey lakes there are eleven islands of different area (Fig. 1). The islands serve as a nesting site for colonial semiaquatic birds including rare ones, such as Relict gull (*Larus relictus*). The islands of Torey section are one of the world's four nesting sites of Relict gull.

The reserve is situated in the zone of extremely continental climate. The peculiar feature of the climate is great amplitude of fluctuations in 24-hour and year temperatures, and uneven distribution of seasonal and many-year precipitation. 24-hour fluctuations of temperature reach 15-20°C, and year ones – 80°C. During dry years 150 mm precipitation fall, in wet years – up to 350 mm. Cyclic changes in humidity with alternation of drought and wet periods about 30 years in duration are characteristic of Dauria [1]. The Torey lakes have a changeable hydrological pattern. On the map charts based on the deciphering of space shots changes in the area of the water surface of the Torey lakes are observed (Fig.2). In the wet period in 2000 the area of the Torey lakes water surface was 852 $km^2$, in 2005 – 750 $km^2$, in 2007 –650 $km^2$, in 2008 – 550 $km^2$, in 2009 г. – 285 $km^2$. The year 2009 was the most

* Corresponding author: bagul72@mail.ru

droughty. In May 2009 Lake Barun-Torey dried up. Lake Zun-Torey got considerably low and the island Aral of the Zun-Torey joined the shore.

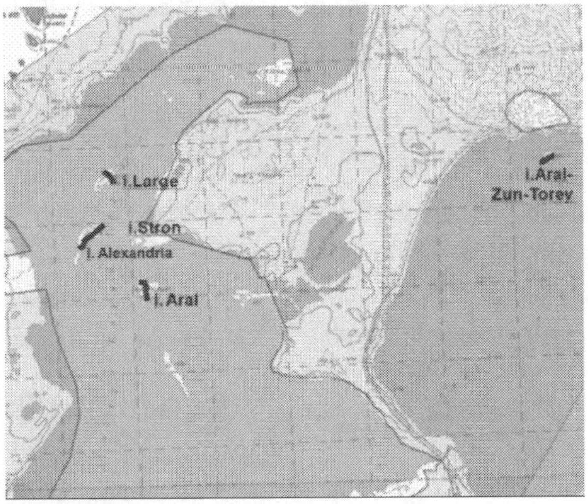


**Fig. 1.** Location of studied islands on Torey lakes

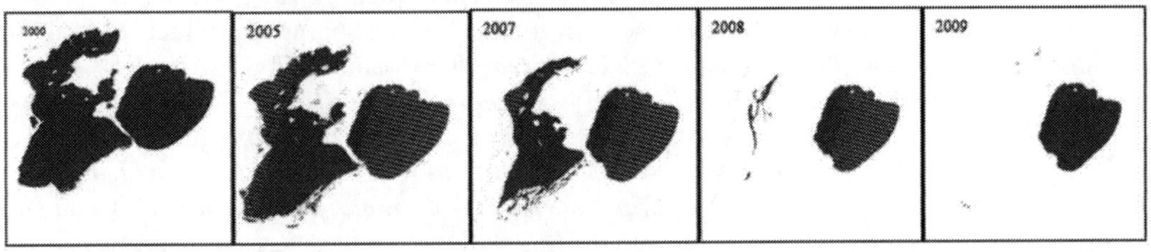


**Fig. 2.** Monitoring of the area of Torey lakes water surface in 2000-2009 rivers Uldza, Imalka that fill the Torey lakes dried up. A lot of small steppe lakes in the Torey hollow ran dry.

## 2 Materials and methods

In 2009 we studied five islands: Alexandria (area – 11 km$^2$), Aral (8.5 km$^2$), Bolshoy (8.7 km$^2$), Kamenny (0.6 km$^2$) and Aral of the Zun-Torey (5.1 km$^2$). On each island we laid a geobotanical profile. Geobotanical descriptions of the vegetation were conducted according to the standard methodology [2]. This paper is based on the material from 78 geobotanical descriptions of the islands vegetation. Classification of the island vegetation on the Torey lakes was done on the ecologo-phytocenotic principles laid out by E.M. Lavrenko [3]. To associations we ascribe communities with a unified composition of edificators and co-edificators, homogeneous synusial structure. Formations are identified by the basic edificator and are joined into groups.

## 3 Results and discussion

The purpose of the study is describing the space structure and diversity of the vegetation on five islands of the Torey lakes in the droughty period.

We considered a typical example of the distribution of island vegetation belts on the geobotanical profile of the Bolshoy island (Fig. 3). The profile has a total length of 1 km 140 m, is oriented from the south-east to north-west and laid across the top of the island (with a height of 609 m above sea level).The spatial distribution of vegetation is shown in Fig. 3 at the level of ecological-phytocenotic classification formations, the description is made at the level of associations. The main influence on the distribution of vegetation has a position of associations in the relief. The relief is generally flat-flat, from the top to the coastal shaft, gently sloping (angle of inclination up to 3 degrees) slopes descend.

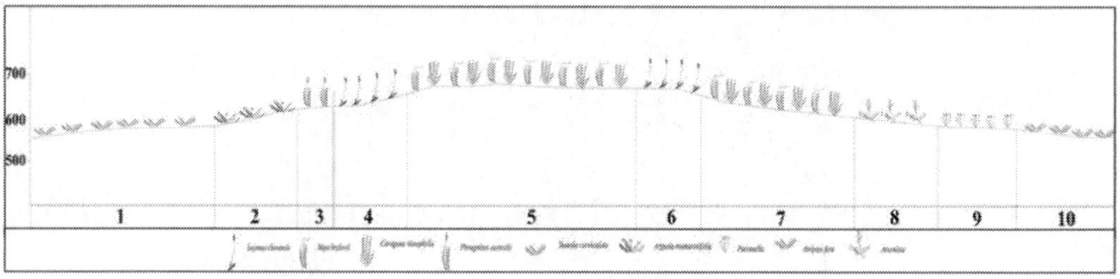


**Fig. 3.** Vegetation of Bolshoy island. Formations: 1, 10. ***Seepweed;*** 2. ***Saltmarsh-grass;*** 3, 9. ***Argusia;*** 4. ***Reed;*** 5, 7. ***Leymus;*** 6, 8 ***Stipa***

On the shores and smooth slopes of the island Bolshoy we distinguish three meadow formations – Argusia, Saltmarsh-grass and Seepweed ones. The Seepweed formation (1, 10) comprises seepweed (***Suaeda corniculata***) and ***Suaeda corniculata – Knorringia sibirica*** associations. The coastal areas of the island, recently freed from the water, are captured by the pioneer halophytic associations from *Suaeda corniculata*. *Puccinellia* formation (2) is represented by the association of the same name, in which the main dominants are *Puccinellia tenuiflora* and *P. macranthera*. The *Argusia* formation (3, 9) includes association ***Argusia rosmarinifolia – Puccinellia tenuiflora***. The upper convex part of the island is occupied by steppe vegetation of *Leymus* and *Stipa* formations. The *Leymus* formation (5, 7) includes association ***Leymus chinensis – Allium polyrhizum***; the *Stipa* formation (6, 8) - association ***Stipa krylovii – Caragana stenophylla***. The reed formation (4), described by us in the middle part of the southeastern slope, consists of hydrophyte monodominant species *Phragmites australis* thickets. Reed thickets in the dry period occupy small areas.

The islands of the Torey Lakes are natural refugiums for rare species and associations. Five rare species have been found on the islands*:* *Asparagus brachyphyllus*, a species listed in the Red Book of the Russian Federation [4] and *Ephedra dahurica*, *Limonium aureum*, *Nitraria sibirica*, *Astragalus miniatus*, species included in the Red Book of the Transbaikal territory [5]. Of the rare and endemic associations, the following were described: ***Oxytropis prostrata – Suaeda corniculata***, ***Oxytropis prostrata – Melilotus suaveolens – Argusia rosmarinifolia*** with the participation of the endemic species *Oxytropis prostrata;* and ***Argusia rosmarinifolia – Convolvulus ammanii***, ***Argusia rosmarinifolia – Puccinellia macranthera – Knorringia sibirica***, ***Artemisia anethifolia – Nitraria sibirica***, ***Argusia rosmarinifolia – Puccinellia macranthera – Puccinellia tenuiflora***, with *Argusia rosmarinifolia* xerophilous tertiary desert-steppe Central Asian relict.

A change of vegetation from halophytic to steppe communities is characteristic of almost all studied islands. An exception is island Kamenny, which has small area – 0.6 $km^2$. In the vegetation of the island, of pioneer seepweed, seepweed-orach, orach associations with participation of annual and biennial halophytic species *Atriplex fera* and *Suaeda corniculata* are described. Steppe vegetation is not developed due to the small area of the island. Hydrophytic vegetation is represented by the Reed formation. Thickets of reeds are noted for the Aral, Alexandria and Bolshoy Islands. In the dry season, reed areas degrade and shrink. Below is the ecologo-phytocenotic classification of the island vegetation of the Torey lakes (Table).

**Table.** Ecologo-phytocenotic classification of the island vegetation of the Torey lakes

| **Vegetation type** | **Name of formation** | **Name of association** |
|---|---|---|
| Steppes | Onion | Sage-onion *(Allium polyrhizum – Artemisia gmelinii – Artemisia frigida)* |
| | | Sage-onion *(Allium polyrhizum – Artemisia anethifolia)* |
| | Sage | Onion-sage (*Artemisia commutata – Allium polyrhizum)* |
| | | Onion-sage *(Artemisia frigida – Artemisia gmelinii – Allium polyrhizum)* |
| | | Caragana-sage *(Artemisia scoparia – Artemisia gmelinii – Caragana stenophylla)* |
| | Wild rye | Sage-onion-wild rye *(Leymus chinensis – Allium polyrhizum – Artemisia scoparia)* |
| | | Stipa-wild rye *(Leymus chinensis – Stipa krylovii)* |
| | | Sage-wild rye *(Leymus chinensis – Artemisia frigida)* |
| | | Caragana-stipa-wild rye *(Leymus chinensis – Stipa krylovii – Caragana stenophylla)* |
| | | Sage-stipa-wild rye *(Leymus chinensis – Stipa krylovii – Artemisia gmelinii)* |
| | | Sage-wild rye *(Leymus chinensis – Artemisia gmelinii – Artemisia scoparia)* |
| | | Onion-sage-wild rye *(Leymus chinensis – Artemisia gmelinii – Artemisia frigida – Allium polyrhizum)* |
| | Stipa | Onion-stipa *( Stipa krylovii – Allium tenuissimum – Allium polyrhizum)* |
| | | Sage-caragana-stipa *(Stipa krylovii – Caragana stenophylla – Artemisia scoparia)* |
| | | Wild rye-stipa *(Stipa krylovii – Leymus chinensis*) |
| | | Sage-stipa *(Stipa krylovii – Artemisia scoparia)* |
| | | Herbs-stipa *(Stipa krylovii – Heteropapus altaicus – Allium polyrhizum)* |
| | Stipa – shrub | Caragana-stipa *(Stipa krylovii – Caragana stenophylla)* |
| | | Sage-caragana-stipa *( Stipa krylovii – Caragana stenophylla – Artemisia scoparia)* |
| | | Caragana-onion-stipa *(Stipa krylovii – Allium polyrhizum – Caragana stenophylla)* |
| | | Nitraria-caragana-sage-stipa steppe *(Stipa krylovii – Artemisia scoparia + Caragana stenophylla – Nitraria sibirica)* |
| | | Caragana-sage-stipa *( Stipa krylovii – Artemisia scoparia + Caragana stenophylla)* |
| | Forbred-leymus-shrubby | Caragana-nitraria-herbs-wild rye *(Leymus chinensis – Thermopsis lanceolata – Saussurea amara + Nitraria sibirica – Caragana stenophylla)* |
| Halophytic meadows | Saltmarsh-grass | Argusia-nitraria-saltmarsh-grass *(Puccinellia macranthera – Nitraria sibirica – Argusia rosmarinifolia)* |
| | | Argusia-saltmarsh-grass *(Puccinellia macranthera – Argusia rosmarinifolia)* |
| | | Knorringia-saltmarsh-grass *(Puccinellia macranthera – Knorringia sibirica)* |
| | | Saltmarsh-grass *(Puccinellia tenuiflora – Puccinellia macranthera)* |
| | | Nitraria–saltmarsh-grass *(Puccinellia macranthera – Nitraria sibirica)* |
| Complex pioneer and | Seepweed | Orach-seepweed *(Suaeda corniculata – Atriplex fera)* |
| | | Seepweed *( Suaeda prostrata – Suaeda corniculata )* |

| Vegetation type | Name of formation | Name of association |
|---|---|---|
| halophytic vegetation | | Argusia-knorringia-seepweed *( Suaeda corniculata – Knorringia sibirica – Argusia rosmarinifolia )* |
| | | Knorringia-argusia-seepweed *( Suaeda prostrata – Argusia rosmarinifolia – Knorringia sibirica )* |
| | | Saltmarsh-grass-seepweed *( Suaeda prostrata – Puccinella macranthera )* |
| | | Pigweed-seepweed *( Suaeda corniculata – Atriplex fera )* |
| | | Knorringia-seepweed *( Suaeda corniculata – Knorringia sibirica )* |
| | | Argusia-saltmarsh-grass-seepweed *( Suaeda corniculata – Puccinellia macranthera – Argusia rosmarinifolia )* |
| | | Argusia-seepweed *( Suaeda prostrata – Argusia rosmarinifolia )* |
| | Kochia | Seepweed-kochia *( Kochia densiflora – Suaeda corniculata )* |
| | | Sage-seepweed-kochia *( Kochia densiflora – Suaeda corniculata – Artemisia comutata )* |
| | Argusia | Bindweed-argusia *( Argusia rosmarinifolia – Convolvulus ammanii )* |
| | | Saltmarsh-grass-argusia *( Argusia rosmarinifolia – Puccinellia macranthera – Puccinellia tenuiflora )* |
| | | Knorringia-saltmarsh-grass-argusia *( Argusia rosmarinifolia – Puccinellia macranthera – Knorringia sibirica )* |
| | | Pigweed-argusia *( Argusia rosmarinifolia – Atriplex fera )* |
| | Orach | Seepweed-orach *(Atriplex fera – Suaeda corniculata)* |
| | | Argusia-orach *( Atriplex fera – Argusia rosmarinifolia )* |
| | | Orach *( Atriplex fera )* |
| | Oxytrope | Seepweed-oxytrope *( Oxytropis prostrata – Suaeda corniculata )* |
| | | Argusia-melilot-oxytrope *( Oxytropis prostrata – Melilotus suaveolens – Argusia rosmarinifolia)* |
| | Melilot | Argusia-melilot *( Melilotus suaveolens – Argusia rosmarinifolia )* |
| | | Knorringia-argusia-melilot *( Melilotus suaveolens – Argusia rosmarinifolia – Knorringia sibirica )* |
| | Sagebrush | Nitraria-sagebrush *( Artemisia anethifolia – Nitraria sibirica )* |
| Hydrophytic vegetation | Reed | Reed *( Phragmites australis )* |

Thus, the analysis of the spatial structure of vegetation showed that the change of plant associations and formations is closely related to the position in the relief: the areas with similar relief have similar vegetation that varies from halophytic to steppe associations. Such a structure of vegetation is characteristic of the large islands of Alexandria, Bolshoi, Aral and Aral-Zun-Torey. As a rule, pioneer and halophytic vegetation grows along the coast and in the coastal zone of the islands. On the gentle slopes of the islands, it is replaced by halophytic saltmarsh-grass meadows, with spots of hydrophytic reed vegetation. On convex summit relief forms, halophytic vegetation is replaced by steppe vegetation. During the dry period, an increase in the area of pioneer halophytic vegetation and a decrease in hydrophytic vegetation were observed on the islands. The diversity of island vegetation of the Torey Lakes is represented by 52 associations belonging to 16 formations and 4 types of vegetation. The islands of the Torey Lakes are natural refugiums for rare species and endemic associations.

## References

1. V. Kirilyuk, T. Tkachuk, O. Kirilyuk, Problems of adaptation to climate change in Dauria river basins: ecological and hydroeconomic aspects, 5 (2012)
2. *Field geobotany*, **3** (L., 1964)
3. E.M. Lavrentyev, *Vegetation of USSR,* **2** (M, 1940)
4. The Red Book of the Russian Federation (plants and fungi) (M, KMK, 2008)
5. Red Book of the Trans-Baikal Territory. Plants (Novosibirsk, House of peace, 2017)

# Activity of Rosaceae Juss. family species in some high-altitudinal plant communities of the Western Sayan and Kuznetsky Alatau (Khakassia)

*Elena* Sazanakova[1*] and *Natalia* Tupitsyna[2]

[1] N. F. Katanov Khakas State University, Lenina str., 90, Abakan, 655000, Russia
[2] V. P. Astafyev Krasnoyarsk State Pedagogical University, Ada Lebedeva str., 89, Krasnoyarsk, 660017, Russia

**Abstract.** In the studied high-altitudinal phytocenoses of the Western Sayan and Kuznetsky Alatau, 22 species and 7 genera of Rosaceae Juss. family are identified, they were found in communities of 3 associations: ***Flavocetrario cucullatae–Dryadetum oxyodontae*** (lichen-dryad tundra), ***Cladonio stellaris–Betuletum rotundifoliae*** (bushy-lichen tundra) and ***Aquilegio glandulosae–Anthoxanthetum odorati*** (alpine low grass meadow). The activity of species of Rosaceae family is the highest in ***Flavocetrario cucullatae–Dryadetum oxyodontae*** association due to *Dryas oxyodonta* (55.90, 3 species), then in descending order ***Cladonio stellaris–Betuletum rotundifoliae*** (48.12, 19 species) and ***Aquilegio glandulosae–Anthoxanthetum odorati*** (40.80, 11 species) go.

## 1. Introduction

Family Rosaceae Juss. includes more than 3.000 species and about 115 genera [1] and is among the leading families in terms of the family spectrum of Khakassia flora [2]. Its representatives are often dominant and subdominant of phytocenoses. The role of the family in the vegetation cover is determined both by the number of species and their activity [3]. According to our data, Rosaceae family in the flora of Khakassia has 102 species and 26 genera [4], which is 3.4% (3.000 species) of the world flora of the family and 29.3% (348 species) of Rosaceae family of Asian Russia [5]. In the flora of the Yenisey Sayans [6], which includes the entire southern part of Khakassia, as well as in the flora of the northwestern part of the Altai-Sayan province [7], which includes the northwestern Khakassia, Rosaceae family is in sixth position, and in some of its areas it reaches higher positions.

Khakassia is a part of the Altai-Sayan mountain region, it includes the western part of the northern macroslope of the Western Sayan, the eastern slope of Kuznetsky Alatauthe Kuznetsk Highlands, the western part of Minusinsk Basin. Khakassia is located at the

* Corresponding author: sazelevik@mail.ru

junction of the Circumboreal and Iran-Turan Regions of the Holarctic Kingdom [8], near the borders of mountain Altai-West-Sayan and Sayan-Baikal floristic provinces [9].

## 2. Materials and methods

Field works were carried out in 2016–2017 by standard technique [10]. 33 complete geobotanical releves were made in the Western Sayan: the ridges of Sailyg-Khem-Taiga, Kokhosh, Monysh, the valley of the Big On river and in Kuznetsky Alatau: Pustaskhyl, Chalbakh-Taskhyl, Kharykh-Taskhyl mountains and the surroundings of Lake Rybnoye. Releves were attributed to 3 syntaxa described before [11-16]. According to eco-floristic approach, the studied high-altitude communities are classified in the following way:
Class ***Carici rupestris–Kobresietea bellardii*** Ohba 1974
Order ***Kobresietalia myosuroides*** Mirkin et al. (1983) 1986
All. ***Dryadion oxyodontae*** Zhitlukhina et Oniščenko ex Chytrý, Pešout et Anenkhonov 1993
Ass. ***Flavocetrario cucullatae–Dryadetum oxyodontae*** Zibzeev, Nedovesova 2014
Class ***Loiseleurio–Vaccinietea*** Eggler ex Schubert 1960
Order ***Betuletalia rotundifoliae*** Mirkin et al. ex Chytrý, Pešout et Anenkhonov 1993
All. ***Empetro–Betulion rotundifoliae*** Zhitlukhina et Oniščenko 1987
Ass. ***Cladonio stellaris–Betuletum rotundifoliae*** Telyatnikov 2015
Class ***Mulgedio–Aconitetea*** Hadač et Klika in Klika et Hadač 1944
Order ***Schulzio crinitae–Aquilegietalia glandulosae*** Ermakov, Shaulo et Maltseva 2000
All. ***Schulzio crinitae–Aquilegion glandulosae*** Ermakov, Shaulo et Maltseva 2000
Ass. ***Aquilegio glandulosae–Anthoxanthetum odorati*** Krasnoborov ex Ermakov et. al. 2000

List of species of every association is considered as coenoflora. The activity of the species of Rosaceae family was calculated in each of three coenoflora using the formula made by L.I. Malyshev and Yu.N. Petrochenko [17] and modified by M.Yu. Telyatnikov [18].

## 3. Results and discussion

*Dryas oxyodonta* Juz. is found in all three types of communities, activity optimum of the species is ass. ***Flavocetrario cucullatae–Dryadetum oxyodontae*** (51.66). In the Western Sayan (Sailig-Khem-Taiga ridges), the coenoses of this association are confined to the altitudes from 2205 m to 2303 m above sea level (alpine belt). They inhabit the tops of mountains in windy areas with a low level of snow cover. In Kuznetsky Alatau (Pustaskhyl mountain), communities are located from 1273 m to 1524 m above sea level, and are present in the upper part of the mountain tundra belt. They inhabit the slopes of the northern exposure. The association is dominated by dwarf shrubs (*Dryas oxyodonta*, *Empetrum nigrum* L., *Vaccinium vitis-idaea* L.) and lichens (*Flavocetraria cucullata* (Bellardi) Kärnefelt et A. Thell, *F. nivalis* (L.) Kärnefelt et A. Thell, *Thamnolia vermicularis* (Sw.) Schaer.). Mosses are negligible (*Aulacomnium turgidum* (Wahlenb.) Schwägr., *Pleurozium schreberi* (Brid.) Mitt., *Rhytidium rugosum* (Hedw.) Kindb.). *Dryas oxyodonta* is often the dominant and edificator, the plant cover of the species is 7–60 %. The average species number is 29 species per 100 sqm.

In the Western Sayan, communities of ass. ***Cladonio stellaris–Betuletum rotundifoliae*** are were studied on the ridges of Sailyg-Khem-Taiga, Kokhosh, Monysh and in the valley of the Big On river. The altitude ranges from 1704 m to 2269 m above sea level (subalpine belt). This association occupies the drained, non-steep mountain slopes (10–20°) of various orientations, and descends in the lower part of the sub-goltsy belt to the valley of the Big On mountain river. In Kuznetsky Alatau (the saddle between Chalbakh-Taskhyl and Kharykh-Taskhyl mountains, near Lake Rybnoy), the association is located in the upper part of the mountain-tundra belt. The altitude ranges from 1.385 m to 1.524 m above sea level. The association occupies the summits of mountains in the areas of temporary streams after snow melting, depressions, zones of fractures, it descends lower down to the subalpine belt. The main dominants are shrubs (*Betula rotundifolia* Spach, *Salix glauca* L., *S. myrtilloides* L.), lichens (*Cladonia rangiferina* (L.) FH Wigg., *C. stellaris* (Opiz) Pouzar et Vezda, *Cetraria islandica* (L.) Ach., *Flavocetraria cucullata*, *F. nivalis*, *Thamnolia vermicularis*) and shrubs (*Empetrum nigrum*, *Vaccinium vitis-idaea*). In this association, *Dryas oxyodonta* is less active (22.04). Species of *Alchemilla* genus (12 species) are widely represented in the Western Sayan in the headwaters of the Big On river; *Alchemilla krylovii* Juz. (4.10) and *Alchemilla sibirica* Zamelis (2.88) have a higher activity. Species *A. anisopoda* Juz., *A. aperta* Juz., *A. bungei* Juz., *A. cryptocaula* Juz., *A. dasyclada* Juz., *A. diglossa* Juz., *A. lipschitzii* Juz., *A. omalophylla* Juz., *A. orbicans* Juz., *A. subcrenata* Buser are found occasionally. The average species number of ass. ***Cladonio stellaris–Betuletum rotundifoliae*** is 38 species per 100 sqm.

Ass. ***Aquilegio glandulosae–Anthoxanthetum odorati*** unites alpine low grass meadows. It is noted in the Western Sayan on the ridges Sailyg-Khem-Taiga, Kokhosh and Monysh. The altitude ranges from 1859 m to 2116 m above sea level. This association is located in the sub-goltsy belt near the snowfields, in the areas of seepage humidification, in small depressions, on steep 15–30° mountain slopes of northeastern orientation. In Kuznetsky Alatau it is noted at the foothills of Pustoskhyl mountain, near Lake Rybnoye. The altitude ranges from 1463 m to 1520 m above sea level. In the alpine low grass meadows, *Dryas oxyodonta* is the least active (0.63), rare and scant species. In this association, species of *Alchemilla* genus are quite active: *Alchemilla dasyclada* (4.47), *A. leiophylla* (4.47), *A. krylovii* (3.46), *A. sauri* (3.46). The average species number of ass. ***A. g.–A. o.*** is 33 species per 100 sqm.

*Sibbaldia procumbens* L. is noted in the same association. In the Western Sayan, this species is common, but scant. The species is common and highly active (18.33) in alpine low grass meadows of Kuznetsky Alatau.

*Dasiphora fruticosa* (Fisch. Ex Lehm.) Juz. is widely represented as part of shrub layer in ass. ***Aquilegio glandulosae–Anthoxanthetum odorati*** (3.46) and ass. ***Cladonio stellaris–Betuletum rotundifoliae*** (3.53) due to frequent occurrence and small plant cover.

22 species and 7 genera of Rosaceae family are identified in the high-altitudinal communities of the Western Sayan and Kuznetsky Alatau. The species activity is shown in table 1.

**Table 1.** The species activity of Rosaceae family of high-altitude communities in the Western Sayan and Kuznetsky Alatau

| Species of the Rosaceae family | Associations | | |
|---|---|---|---|
| | *F. c. – D. o.* | *C. s. – B. r.* | *A. g. – A. o.* |
| *Alchemilla anisopoda* | – | 1.33 | 0.63 |
| *Alchemilla aperta* | – | 0.77 | – |
| *Alchemilla bungei* | – | 0.24 | – |
| *Alchemilla cryptocaula* | – | 0.24 | – |
| *Alchemilla dasyclada* | – | 0.24 | **4.47** |
| *Alchemilla diglossa* | – | 0.24 | – |
| *Alchemilla krylovii* | – | **4.10** | **3.46** |
| *Alchemilla leiophylla* | – | – | **4.47** |
| *Alchemilla lipschitzii* | – | 0.77 | – |
| *Alchemilla omalophylla* | – | 0.24 | 0.63 |

| | | | |
|---|---|---|---|
| *Alchemilla orbicans* | – | 0.77 | – |
| *Alchemilla sauri* | – | – | **3.46** |
| *Alchemilla sibirica* | – | 2.88 | 0.63 |
| *Alchemilla subcrenata* | – | 0.24 | 0.63 |
| *Cotoneaster uniflorus* | – | 0.24 | – |
| *Dasiphora fruticosa* | – | **3.53** | 3.46 |
| *Dryas oxyodonta* | **51.66** | **22.04** | 0.63 |
| *Potentilla evestita* | – | 2.18 | – |
| *Potentilla gelida* | 0.77 | **5.15** | – |
| *Potentilla nivea* | **3.47** | **–** | **–** |
| *Sibbaldia procumbens* | – | 2.43 | **18.33** |
| *Spiraea alpina* | – | 0.49 | – |
| Total: | 55.90 / 3 species | 48.12 / 19 species | 40.80 / 11 species |

## 4. Conclusion

The activity of species of the Rosaceae family is highest in lichen-dryad tundra ass. ***Flavocetrario cucullatae–Dryadetum oxyodontae*** – 55.90 (3 species), in bushy-lichen tundra ass. ***Cladonio stellaris–Betuletum rotundifoliae*** – 48.12 (19 species), in alpine low grass meadows ass. ***Aquilegio glandulosae–Anthoxanthetum odorati*** – 40.80 (11 species).

This paper was supported by a grant No 18-44-24006 “Natural and urbanized floras of Trans-Yenisei Siberia” given by the Russian Foundation for Basic Research and the Krasnoyarsk Regional Supporting Fund of Scientific and Technical Activities.

## References

1. R.V. Kamelin, *Rosaceae* (Barnaul, Altaiskye Stranitsy, 2006)
2. *The vegetation cover of the Khakassia* (Novosibirsk, Science, 1976)
3. B.A. Yurtsev, *Flora Suntar-Hayat*: *Problems of the history of high-altitude landscapes of North-Eastern Siberia* (Leningrad, Science, 1968)
4. E.V. Sazanakova, N.N. Tupitsyna, Vestnik KrasGAU, **4** (115) (2016)
5. *Synopsis of flora of Asiatic Russia: Vascular plants* (Novosibirsk, SD RAS, 2012)
6. N.V. Stepanov, *Vascular plants of the Yenisei Sayan* (Krasnoyarsk, Sib. Fed. University, 2016)
7. A.L. Ebel, *Synopsis flora north-western part of the Altai-Sayan province* (Kemerovo, KREOO «Irbis», 2012)
8. A.L. Takhtajan, *The floristic regions of the World* (L., Science, 1978)
9. R.V. Kamelin, *Problems of botany of South Siberia and Mongolia* (Barnaul, AzBuka press, 2002)
10. A.A. Yunnatov, *Field geobotany*. **3** (M.-L., Science, 1964)
11. E. G. Zibzeev, Sib. Bot. journal, **2**, 2 (2007)
12. N.B. Ermakov, *The current status of the basic scientific concepts of vegetation* (Ufa, Gilem 2012)
13. E.G. Zibzeev, E.A. Basargin, Herald of Novosibirsk State University. Series Biology, clinical medicine, **10**, 2 (2012)
14. E.G. Zibzeev, T.A. Nedovesova, Turczaninowia, **17**, 3 (2014)
15. M.Yu. Telyatnikov, Flora of Asian Russia, 3 (19) (2015)

16. E.A. Basargin, E.G. Zibzeev, Flora of Asian Russia, 1 (29) (2018)
17. L.I. Malyshev, Bot. Journal, **58**, 11 (1973)
18. M.Yu. Telyatnikov, Siberian Ecological Journal, 6 (2010)

# The influence of willow plantations and soil composition on the process of natural vegetation settlement at oil drilling sites in Western Siberia

*Vladimir* N. Sedykh[1], *Vyacheslav* V.Tarakanov[1*], and *Mikhail* Yu.Telyatnikov[2]

[1]West-Siberian Branch of the Sukachev Institute of Forest SB RAS - Branch of the Federal Research Center "Krasnoyarsk Science Center", Zhukovsky Str., 100/1, Novosibirsk, 630082, Russia
[2]Central Siberian Botanical Garden SB RAS of the SB RAS, Zolotodolinskaya Str., 101, Novosibirsk, 630090, Russia

**Abstract.** Test cultures of various willow species were created in 1997 on a technogenic sandy site of one of oil deposits (62°46'18.36" N, 72°09'48.69" E). Sixty-five isolated plots, including various soil substrates (sand, peat, drilling wastes), were placed in the testing area of 684 m$^2$. Optimum combinations of substrate components were identified, and the tolerance of willow species to drilling wastes was revealed by the results of investigating 1-5-year cultures [1]. In 2016 we investigated the natural settlement of aborigine plants in this test plot. The willow cultures mortality was high according to short offspring vegetative lifetimes and the limitation of feed areas. However, willow cultures and soil reclamation stimulated a settlement of natural plant seedlings. The comparison of phytocenoses, which were formed in various experimental and "background" conditions, has allowed making some conclusions. 44 plant species are growing now in the experimental area. The least number of species is characteristic for pure sand and pure peat substrates, the greatest - for complex substrates with sand, peat and drilling wastes. Thus, the experimental willow cultures and partial soil reclamation accelerated the formation process of forest ecosystems that created a "plant oasis" on a deserted landscape of technogenic sand.

## 1 Introduction

The north of Western Siberia resembles the arena of continuous "struggle" between forests and marshes [2]. At present, as a result of the activities of the oil and gas complex, dry-bottom habitats of sand and peat have been created on the site of a significant part of the former wetlands [1]. Gradually, they turn into "man-made" forests. But without human assistance, this process can take several decades [3].

The study of the dynamics of vegetation appearance in the technogenic landscapes of the oil complex shows that, at the initial stage of the forest formation process, the "pioneer"

* Corresponding author: tarh012@mail.ru

species of genera *Salix, Populus* and *Betula* play an active role. They create a forest environment and contribute to the settlement of other species, including conifers.

With this in mind, the forest reclamation technology, which "launches" the mechanism of the initial stage of forest formation by creating cultures from local willow species [4], has been developed. To study the reaction of these species to peat-sand substrates contaminated with drilling wastes, a test site for forest reclamation was created about 20 years ago [5].

The studies carried out in 1–5 years after plantations allowed for the selection of reclamation- and recultivation-promising willow species and soil compositions, respectively. However, the issue of the extent to which artificial plantations and sand reclamations contribute to the process of natural overgrowing of this experimental area have not yet been studied.

The purpose of this work is to assess the productivity and species diversity of the phytocenosis, which was formed at the experimental site after its recultivation.

## 2 Materials and Methods

The experimental forest reclamation site with the total area of 684 $m^2$ was created on the unexploited sandy part of the well drilling site in the Tyansk (now Muryaunsky) oil field of OJSC "Surgutneftegaz" located on the territory of ridge-bog complex in the middle taiga zone (62° 4618.36" n. lt. 72° 09'48.69" e. lng.) during1997–1998. The methodical features of the experiment are described in the monograph [1].

Experimental plots with the total area of 261 $m^2$ included 61 box sections, each with the area of 1 $m^2$ and depth of 0.3 m, as well as 4 micro-grounds, each with the area of 50 $m^2$ and depth of 0.5 m (Fig. 1). They were intended to study the dose-effect relation in a wide range of ratios of drilling waste, peat and sand, and the concentration of each of which varied from 0 to 80-100 %, also for testing various species that are *a priori* promising for forest reclamation. At the same time, most of the grounds were planted by *Salix triandra* L., *S. dasyclados* Wimm., *S. viminalis* L., *S. pentandra* L. and *S. bebbiana* Sarg. Also, wild species of *Betula pendula* Roth., Ehrh ., *Sorbus sibirica* L., seedlings of *Larix sibirica* Ledeb. and hybrid forms of the poplar breeding of CSBG SB RAS [6] were planted in small numbers. In addition, *Pinus sylvestris* L., *Larix sibirica* Ledeb. and *Hippophae rhamnoides* L. seeds were sown in plots of 1-2 $m^2$.

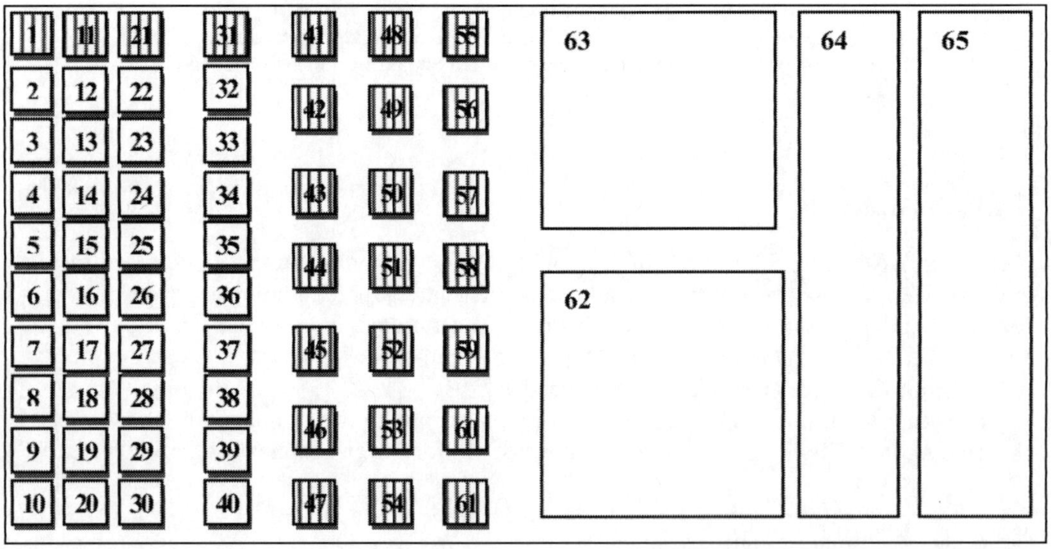

**Fig.1.** The layout of the plots within the recultivated site. Plots № 1-30 - box sections (each of 1 m$^2$) with substrates of sand and drilling waste; No. 31-40 - box sections of 1 m$^2$ with substrates of sand and peat, to which the drilling waste was added in 2004 and which were temporarily used for growing vegetables; No. 41-61 - box sections of 1 m$^2$ with substrates of sand, peat and drilling waste; No. 62 - section of 50 m$^2$ with a sand substrate; No. 63 – section of 50 m$^2$ with a substrate of sand and drilling waste; No. 64 – section of 50 m$^2$ with a peat substrate; No. 65 – section of 50 m$^2$ with a substrate of sand and peat.

At the end of August, 2016, in the experimental plots and on the land free from plantations, the list and abundance of plant species and tree-stand taxation parameters were estimated using generally accepted geobotanical and silvicultural methods.

## 3 Results and discussion

In order to create artificial phytocenoses, 12 woody plant species, with over 90 % of the plots occupied by 5 willow species, were involved. Forty-four plant species, unevenly distributed among the experimental variants (Tables 1, 2), were found in the recultivated area in twenty years after the planting.

Of these, 5 tree species and 5 willow species were preserved from the original plantation. But if birches and pines increased their share of participation in the phytocenosis due to the growth of saplings and the emergence of self-sowing, the aging willow cultures are at an older age, and their participation in the phytocenoses decreased to 0-38 %, being 15% on the average (Table 3). The number of species increased, in comparison with the plantation, by 34. This was mainly due to the species of herbaceous plants, mosses and shrubs (Table 2).

**Table 1.** List of plant species found in 2016 within the recultivated area

| Storey | List of plant species | N |
|---|---|---|
| Trees | *Larix sibirica, Pinus sylvestris, Populus nigra, P. hybryda, Betula pendula* | 5 |
| Tall shrubs | *Sorbus sibirica, Salix bebbiana, S. dasyclados, S. pentandra, S. triandra, S. viminalis* | 6 |
| Shrubs | *Betula nana* | 1 |
| Small shrubs | *Empetrum nigrum, Ledum palustre, Vaccinium myrtillus, V. uliginosum, V. vitis-idaea* | 5 |
| Herbs | *Calamagrostis langsdorffii, Carex species, C.aquatilis, C. limosa, Chamaenerion angustifolium, Dactylorhiza fuchsii, Diphasiastrum complanatum, Drosera rotundifolia, Equisetum arvense, Eriophorum polystachion, E. vaginatum, Juncus compressus, Oxycoccus palustris, Poa palustris, Rubus chamaemorus, Stellaria palustris, Trifolium pratense, Vicia cracca.* | 18 |
| Mosses | *Aulacomnium turgidum, Sphagnum species, Polytrichum piliferum, P. strictum, P. species, Pohlia sp., Hypnum mosses* | 7 |
| Lichens | *Cladonia species, C. coccifera* | 2 |
| Total | | 44 |

**Table 2.** Number of plant species (psc.) in different recultivated plot variants

| Storey | Box sections (each of 1 m$^2$) | | | Micro-grounds (each of 50 m$^2$) | | | | Sand interlocking areas |
|---|---|---|---|---|---|---|---|---|
| | S+DW (№ 1-30) | S+DW+P (№ 31-40) | S+DW+P (№ 41-61) | S (№ 62) | S+DW (№ 63) | P (№ 64) | S+P (№ 65) | |
| Trees | 1 | 4 | 3 | 4 | 2 | 1 | 1 | 5 |

| | | | | | | | | |
|---|---|---|---|---|---|---|---|---|
| Tall shrubs | 3 | 4 | 4 | 4 | 1 | 0 | 0 | 1 |
| Shrubs | 1 | 1 | 1 | 0 | 1 | 1 | 1 | 0 |
| Small shrubs | 3 | 6 | 3 | 3 | 6 | 5 | 5 | 5 |
| Herbs | 8 | 15 | 11 | 1 | 5 | 5 | 5 | 3 |
| Mosses | 2 | 6 | 6 | 1 | 3 | 2 | 2 | 4 |
| Lichens | 0 | 1 | 2 | 1 | 1 | 0 | 0 | 1 |
| Total | 17 | 36 | 29 | 14 | 18 | 13 | 13 | 19 |

**Note**: S, DW, P – sand, drilling waste and peat, respectively.

If we compare different experimental variants with each other, as well as experimental plots with areas free from plantations, we can note the following. The largest number of plant species is in the range of 29-36 pcs., and the largest participation of woody plants with a total projective cover of 37–42 % is characteristic for variants with a complex composition of substrates which include sand, peat and drilling waste (see Table 2-3). At the same time, the variant of temporary use of plots for growing vegetables (sections 31-40) also differs by the maximum number of herbaceous plant species, which could be caused by their additional reclamation (due to the introduction of additional biomass into the soil and its loosening). On the average, the smallest projective cover of 0–28 % vegetation is observed on the sandy ground (landfill) No. 62 and in the sandy areas free from experimental plantations.

**Table 3** Total projective cover by plants (%) in different recultivated plot variants

| Storey | Box sections (each of 1 m$^2$) | | | Micro-grounds (each of 50 m$^2$) | | | | Sand interlocking areas |
|---|---|---|---|---|---|---|---|---|
| | S+DW (№ 1-30) | S+DW+P (№ 31-40) | S+DW+P (№ 41-61) | S (№ 62) | S+DW (№ 63) | P (№ 64) | S+P (№ 65) | |
| Trees | 19 | 43 | 37 | 16 | 22 | 15 | 22 | 24 |
| Tall shrubs | 38 | 3 | 28 | 5 | 3 | 0 | 21 | 0 |
| Shrubs | 3 | 2 | 1 | 0 | 2 | 4 | 5 | 0 |
| Small shrubs | 2 | 4 | 3 | 2 | 8 | 11 | 10 | 2 |
| Herbs | 18 | 15 | 11 | 2 | 8 | 13 | 9 | 2 |
| Mosses | 58 | 25 | 29 | 6 | 24 | 52 | 17 | 28 |
| Lichens | 0 | 1 | 1 | 1 | 1 | 0 | 1 | 3 |
| Total | 132 | 61 | 88 | 26 | 61 | 136 | 101 | 36 |

**Note**: S, DW, P – sand, drilling waste and peat, respectively.

The integral taxation-botanical characteristics of groups of variants suggests that the amelioration of poor sandy substrates with peats and drilling waste, in combination with woody plantations, first of all willow species, leads to the accelerated formation of closed birch-willow stands with an admixture of conifers (Table 4) . Favorable conditions for the formation of tree phytocenoses are also created in the interplot space, where the undergrowth thickness 38 thousand pcs/ha. At the same time, at a distance from the recultivated area, the sandy pad areas are practically not overgrown.

**Table 4.** Taxation-botanical characteristics of plantations (phytocenoses) in the recultivated area (RA)

| Experimental variant | Stand composition | Stand density, psc/ha. | Stand stock, $m^3$/ha. | Undergrowth density, th.psc./ha. | Total of plant species, psc. | Total projective cover, % |
|---|---|---|---|---|---|---|
| Sections of 1 $m^2$ | 8S2B +P | 16143 | 21 | 49.4 | 31 | 116.0 |
| Sections of 50 $m^2$ | 6B3L1S+P | 12150 | 39 | 29.7 | 31 | 83.8 |
| Interplot areas | 5B3S2P | 7785 | 15 | 38.3 | 21 | 36.1 |
| Area adjacent to the RA | – | – | – | 11.4 | - | <5.0 |

## 4 Conclusions

Creating cultures of willows and other woody plant species from "pioneer" ones, also the "mosaic" melioration of sands, accelerate the process of oil well drilling sites natural overgrowth in Surgut Woodland. In the future, close species-rich birch-willow stands with an admixture of conifers may form in such areas.

## References

1. M.I. Neyshtadt, *Scientific prerequisites of developing swamps of West Siberia* (M, Nauka, 1977)
2. V.N. Sedykh, V.V. Tarakanov. *Woody plant tolerance to drilling wastes* (Ed. Miljutin, 2004)
3. M.N. Gasheva, A.V. Soromotin, S.N. Gashev, Ecology, **2** (1990)
4. Yu.N. Ilichev, V.V. Tarakanov, I.A. Galkin. Forestry, **5** (2002)
5. V.N. Sedykh, V.V. Tarakanov, V.T. Bakulin. Forest science, **3** (2003)
6. V.T. Bakulin. The introduction and breeding of poplar in Siberia (Novosibirsk, 1990)

# The ontogeny and the demographic structure of the coenopopulations of the species *Oxytropis chakassiensis (Fabacaea)* in the steppe communities of Khakhassia (Russian Federation)

*Inessa* Selyutina[1*], *Svetlana* Lebedeva[2], *Evgeny* Zibzeev[1]

[1]Central Siberian Botanical Garden SB RAS Zolotodolinskaya Str., 101, Novosibirsk, Russia
[2]State Reserve "Khakassky", Tsukanova Str., 164, Abakan, Russia.

**Abstract.** The article presents the results of the analysis of the demographic structure of ten coenotic populations of *Oxytropis chakassiensis* in near-Yenissei steppes (Khakassia and Krasnoyarsk krai). Most part of studied species coenopopulations are not complete and definitive, normal, mature, have a bimodal developmental spectrum with peaks at g1 (v)- and g3-individuals. Changes in the structure of the ontogenetic spectrum (increasing the share of young plants) depend mainly on the environmental conditions and the degree of human disturbance of habitats: rocky slopes and decrease in pasture digression create more favorable conditions for the survival of juveniles. The ontogeny of *O. chakassiensis* (the rare species of near-Yenissei steppes) was studied and its life- form was described. Four periods and nine ontogenic stages were distinguished. The results of our study showed that *O. chakassiensis*is was a petrophyte steppe species with strict ecological and cenotic propensity to petrophytic steppe with the prevalence *Koeleria cristata*, *Festuca valesiaca*, *Arctogeron gramineum*, *Hedysarum turczaninovii*, *Alyssum obovatum* and *Thymus minussinensis*. Due to long generative period *O. chakassiensis* is successfully renewed and is stable during long periods of extreme conditions of petrophytic steppes with high pasture digression.

It is well-known that, in order to evaluate the state and the possibility of preservation of rare species, it is necessary to study their biological characteristics and the structure of their populations. Yu. A. Zlobin [1] identified the study of the populations of the rare, endangered and protected species as one of the major tasks of the population ecology, with the purpose of developing of effective methods of maintaining their stable existence. The use of the population and ontogeny assessment approach is required to evaluate the state of species in plant communities [2, 3].

*O. chakassiensis* Polozhij is a rare species, endemic of the near-Yenissei steppes (the south of Middle Siberia), which is registered in the Red Books of the Republic of

* Corresponding author: selyutina.inessa@mail.ru, egzibzeev@gmail.com

Khakassia [4] and of the Krasnoyarsk krai [5]. It is protected in the Khakasskiy reserve in the areas of Lake Itkul and Lake Bele. Outside the reserve, the species is found in steppe communities, which are being intensely used in farming and are exposed to pasture of cattle and to recreational damage. The ecological and coenotic optima of *O. chakassiensis* are within a narrow range; therefore, this species is primarily endangered.

The ontogeny and the demographic structure of ten (10) coenotic populations (CP) of *O. chakassiensis* were studied in the steppe regions of the Republic of Khakassia. The studies were carried out on the territories of the reserve clusters of Lake Itkul and Lake Bele and outside the reserve. The ontogenetic states were identified according the concept of discrete description of ontogeny [3, 6, 7, 8, 9]. The life form (biomorph) was determined by studying the adult plants of the middle generative age [10, 11]. The ontogenetic structure of the coenopopulations were analyzed using delta-omega ($\Delta$-$\omega$) criterion proposed by L.A. Zhivotovsky [12]. The effective [12] and ecological density of plants in the coenopopulations were determined [13]. A plant was a unit of study. For integrated characteristics of the population structure, the following demographic characteristics were used: $\Delta$ – the age index [8], $\omega$ – the effectiveness index [12], the recovery index ($I_r$) and the ageing index – $I_a$ [14].

*O. chakassiensis* belongs to *Xerobia* section. The area of this endemic species of the Yenissei steppes (Abakan, Iyus-Shira and Minusinsk regions) is small, and the species is known to exist only in 5 locations: near the Charkova ulus and the Kapchala station of the Ust-Abakan district, Lake Itkul, Lake Bele and Lake Vlasyevo in the Shira district and in the vicinity of Lake Uchum in the Krasnoyarsk krai [4, 5]. *O. chakassiensis* grows on flat hill tops or on southern and south-eastern hill slopes, more rarely, on south-western and western hill slopes up to 12° steep, within the range of altitudes varying from 430 to 650 a. s. l. The species is found among petrophytes of typical steppes, their cover range from 40 to 90%. The projective cover of *O. chakassiensis* is small, mostly less than 1%, and much more rarely varies from 1% to 3%. *O. chakassiensis* is found in low-grass steppes, where bunch grasses (*Koeleria cristata* (L.) Pers., *Festuca valesiaca* Gaudin), as well as petrophytic low herbs (*Arctogeron gramineum* (L.) DC., *Hedysarum turczaninovii* Peschkova, *Alyssum obovatum* (C.A. Meyer), *Thymus minussinensis* Serg.) dominate.

*O. chakassiensis* is a taproot polycarpic hemicryptophyte herb with a pauciramose immersed caudex, vegetative polycyclic rosette shoots and elongated monocarpic generative axil shoots.

Multicipital perennial taproot herbs with polycarpic shoots of a rosette type are biomorphs of a monocentric type, incapable for vegetative reproduction. They reproduce only by seeds.

To identify the ontogenetic states of the plants, we used the following features: the presence of cotyledons, the structure of leaves and their number, formation of a caudex, the ratio of the processes of growth and destruction and the number of vegetative and generative shoots. Based on these features, 4 ontogenetic periods and 9 ontogenetic states were identified (Fig.1).

**The stage of latency.** The seeds are rounded and have the shape of a bud, they are brown colour, about 2 mm long and 1.5 mm wide. The fruit is unilocular, has a rounded shape tapering on top.

**The pre-generative stage.**

Sprouts. For *Oxytropis chakassiensis*, above-ground germination of seeds is characteristic. Sprouts have the main rosette shoot with two rounded cotyledons (3-4 mm long and 5-7 mm wide) and a root 3-4 mm long. The first true leaves are simple, ellipsoid, plain, 8-15 mm long and 2-4 mm wide.

The juvenile adult state. The plants in this state have one rosette shoot 1,5-1,7 cm tall, with one-three simple leaves of a lanceolate or ellipsoid shape. The length of the main root is about 8 cm.

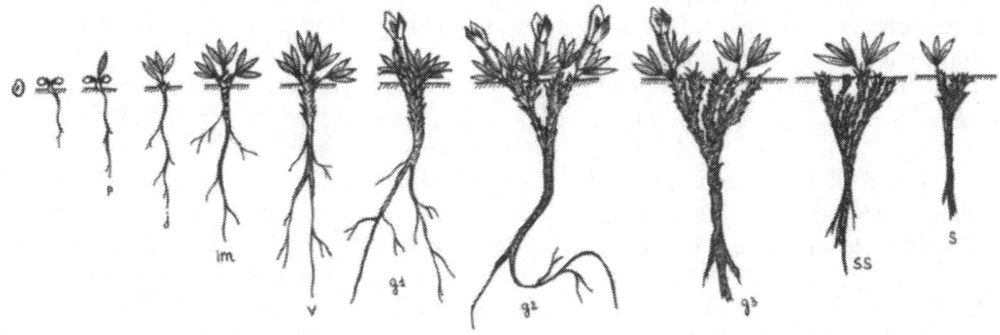


**Fig. 1.** The ontogeny of *Oxytropis chakassiensis*.

The immature state. Plants have the primary mopodially growing rosette shoot with 2-4 simple and ternate parted leaves, they are 1.5-2 cm long and 1-1.2 cm wide, the stalks of them are 1-1.5 cm long. In immature plants, lateral buds are formed in the leaf axils; a caudex begins to be formed, its main root reaches 11cm. The virginal state. The virginal plants have one or, more rarely, two vegetative rosette shoots with 2-5 ternate leaves of an adult type up to 1.8 cm long. The plants develop a multicipital caudex, the taproot of which reaches 18 cm.

**The generative stage.** Young generative plants have one or two monopodial vegetative shoots, with 2-5 ternate leaves on the shoot; the length of a leaf varies from 1.2 to 2 cm. 1-4 generative shoots are formed in a plant. The caudex is one- or two-headed, up to 1-3 cm in diameter. The length of the taproot reaches 20 cm. Middle-aged generative plants have a well-developed multicipital caudex, from 3 to 11 cm in diameter, on which 3-6 vegetative rosette shoots and up to 12 generative shoots are formed. From 2 to 6 leaves, the mean length of which is 2.1±0.2 cm, grow on the vegetative shoot. Old generative plants are characterized by 5-6 vegetative shoots and 2-4 generative shoots. The caudex is divided into individual particles, up to 50 of which are dead particles.

**The postgenerative stage.** The subsenile age state. Subsenile plants have vegetative rosette shoots with 1-2 leaves located on the periphery of the caudex, of an adult and immature types. Senile plants occur extremely rarely and are represented by a single vegetative shoot with leaves of immature or juvenile types.

The study of the ontogenic structure of the plants revealed that all studied coenopopulations of *O. chakassiensis* are normal, definitive, and in most cases, incomplete (7 incomplete CP out of 10 studied populations). Plants in a senile state were absent most often, followed by those in the pregenerative state. For the coenopopulations of the studied species, the domination of dextral ontogenetic spectra is characteristic, with the absolute maximum in g3-plants, in six CP. The coenopopulations with sinistral and centered spectra (two of each type) were equally represented. Based on Zhivotovsky's classification, the coenopopulations of the studied species mostly refer to mature (60 % CP) and ageing 18 %) types, and the young and transitional types characterize one coenopopulation each.

For all studied CP, significant dominance of the share of the generative plants is characteristic, 40.3-96.5 % (Table 1). The fraction of pregenerative plants varies rather much, primarily due to the pressure of pasture and secondly, due to the environmental and coenotic situation. The fraction of plants at the pregenerative stage varies within 2.6–52.2 %, but in most cases, the fraction of the young plants is low and constitutes 2.6–14.9 %. The percentage of virginal plants ensuring continuous replenishment of the generative fraction is low too (2.7 – 8.6 %), and in two CPs virginal plants are absent. The values of the recovery index ($I_R$) vary from 0.03 to 0.12. This indicates insufficiently favourable conditions for recovery of *O. chakassiensis* in its populations. The share of the postgenerative plants in the ontogenic ranges of *O. chakassiensis* varies from 2.7 to 13 %. The plant density in the coenopopulations of *O. chakassiensis* fluctuates from 1.6 to 7 plants/m$^2$, with the values of effective density (density of generative plants) varying from 0.2 to 5.1 plants per 1m$^2$.

**Table**. Demographic characteristics of coenotic populations of *Oxytropis chakassiensis*

| Name of coenotic populations | Plant density | | Absolute and local maximums | Ontogenetic groups, (%) | | | Δ | ω | Ir | Ia |
|---|---|---|---|---|---|---|---|---|---|---|
| | individuals/ m2 | generative individuals/ m2 | | j- v | g1 - g3 | ss, s | | | | |
| Vlasyevo | 3,4 | 2,6 | g1 , g3 | 11,8 | 85,3 | 2,9 | 0,53 | 0,76 | 0,12 | 0,03 |
| Malyy Kobezhikov | 2,0 | 1,5 | g1, g3 | 3,1 | 86,1 | 10,8 | 0,63 | 0,75 | 0,03 | 0,11 |
| Tus | 0,3 | 0,2 | g3 | 2,6 | 84,2 | 13,2 | 0,63 | 0,76 | 0,03 | 0,13 |
| Chalpan | 4,9 | 3,8 | g2 | 14,9 | 82,4 | 2,7 | 0,41 | 0,78 | 0,15 | 0,03 |
| Itkul 1 | 3,3 | 2,2 | g1 , g3 | 21,4 | 68,6 | 10,0 | 0,42 | 0,66 | 0,24 | 0,1 |
| Itkul 2 | 4,8 | 2,2 | im, g1 | 52,2 | 40,3 | 7,5 | 0,23 | 0,46 | 0,56 | 0,07 |
| Itkul 3 | 4,5 | 3,3 | g1, g3 | 11,1 | 78,8 | 10,1 | 0,51 | 0,74 | 0,12 | 0,10 |
| Itkul 4 | 7,0 | 5,1 | g1, g3 | 7,2 | 84,8 | 8,0 | 0,53 | 0,73 | 0,08 | 0,08 |
| Itkul 5 | 1,6 | 1,3 | g2 | 0 | 91,0 | 9,0 | 0,50 | 0,83 | 0 | 0,09 |
| Itkul 6 | 2,0 | 1,6 | g1, g3 | 3,5 | 96,5 | 0 | 0,54 | 0,82 | 0,04 | 0 |

**Note**: Δ – the age index [8], ω – the effectiveness index [12], the recovery index ($I_r$) and the ageing index – $I_a$ [14].

The basic (most common) ontogenetic range for this species is dextral with the absolute maximum in g3-plants and the local maximum in g1-plants. Plant density in the coenopopulations of *O. chakassiensis* ranges within 1.6 to 7 plants/m$^2$, with the values of effective density varying from 0.2 до 5.1 plants per 1m$^2$.

The conducted studies led us to conclude that the ontogeny of *O. chakassiensis* is simple, complete (p-s), and remains permanent in all habitats. The ontogenetic structure of most part of studied coenopopulations is characterized by a large share of generative plants and, in combination with a lengthy generation period, creates a possibility of regular regeneration and sustained existence of populations. Evaluation of the status of the coenopopulations has shown that the condition of studied species is stable and that moderate recreational load does not negatively affect the structure of its coenopopulations. As the anthropogenic load increases, the viability of coenopopulations goes down, the plant density decreases, the ontogenetic ranges become incomplete and dextral, and the populations acquire regressive traits.

Based on the conducted studies, we may conclude that, due to the scantiness and the small number of plants in them, the small share of pregenerative plants and the prevalence of g3-plants, the coenopopulations of *O. chakassiensis* need more thorough protection. *O. chakassiensis* should be recommended to be included into the Red Book of Endangered Species of the Russian Federation.

The study was performed as a state assignment for the Central Siberian Botanical Gardens SB RAS) (state registration number AAAA-A17-117012610052-2), as well as with the financial support from the Russian Foundation for Basic Research (project № 17-04-00076).

## References

1. Yu.A. Zlobin, *Population ecology of plants: current state, growth points* (Sumy, 2009)
2. O.V. Smirnova, Bull. MOIP. Otd. Biol. **85**, 5 (1980)
3. *Cenopopulations of plants (essays on population biology)* (M, 1988)
4. *Red book of the Republic of Khakassia* (Novosibirsk, 2012)
5. *Red book of Krasnoyarsk region* (Krasnoyarsk , 2012)
6. T.A. Rabotnov, *The Life cycle of perennial herbaceous plants in meadow coenoses*, Tr. Bot. Inst. Akad. Nauk SSSR:Geobotanika (M.-L., 1950)
7. A.A. Uranov, *Ontogenesis and age composition of cenopopulations of flowering plants* (M, Nauka, 1967)
8. A.A. Uranov, Scientific. Doc. higher. schools. Biol. sciences, 2 (1975)
9. *Cenopopulations of plants (basic concepts and structure)* (M, 1976)
10. I.G. Serebryakov, *Polevaya geobotanika,* **3** (Moscow, Nauka, 1964)
11. T.I. Serebryakova, Results of science and technology. Ser. Botany. VINITI, **1** (1972)
12. L.A. Zhivotovsky, J. Ecology, 1 (2001)
13. Yu. Odum, *Ecology*, **2** (M. 1986)

1\. N.V. Glotov, On the estimation of age structure parameters of plant populations in Plant life in heterogeneous environment, **1** (Yoshkar-Ola, 1998)

# Materials on the distribution of plant species which are rare for the Volgograd oblast

*Stepan* Senator[1*], *Vladimir* Vasjukov[1], *Eugene* Zibzeev[2], and *Andrey* Korolyuk[2]

[1]Institute of Ecology of the Volga river basin of RAS, 445003, Komzina Str., 10, Togliatti, Russia
[2]Central Siberian Botanical Garden, SB RAS, 630090, Zolotodolinskaja Str., 101, Novosibirsk, Russia

**Abstract.** The information about the distribution of 28 rare and endangered species of vascular plants under Federal and regional security, registered in 2018 on the territory of the Volgograd oblast under the Federal and regional security is given. Among them, 2 species (*Crambe tataria* Sebeók и *Stipa zalesskii* Wilensky ex P. A. Smirn.) are included in the List of Threatened plant species of Europe. Each location is linked to the grid of Atlas Flora Europaea. The obtained materials about the distribution of rare and endangered species of vascular plants represent an important data for long-term monitoring and also for environmental management measures at the regional level.

## 1. Introduction

Volgograd oblast is located in the South-East of the European part of Russia in the basins of the Lower Volga and the Middle Don. The majority of this region belongs to the steppe zone. Zonal vegetation is herb-bunchgrass (moderately arid) and bunchgrass (dry) steppes, which are now replaced by half-shrub-bunchgrass desert steppe to the East and South-East. Only small area in the North-West belongs to the forest-steppe zone.

This region has high, unique for lowlands vascular plant diversity: more than 2970 species were registered there [1]. In this regard, information on the distribution of rare and threatened with extinction species is valuable from a scientific and environmental point of view. The present work provides new information, specifying the distribution of some vascular plants under the protection at the regional and Federal levels.

## 2. Materials and methods

Field studies were implemented May 22–31, 2018 in the Volga region and the Middle Don. It is the territory of the Frolovsky, Gorodishchensky, Ilovlinsky, Kalachevsky, Leninsky, Nikolaevsky and Pallasovsky districts of the Volgograd oblast.

The following sources were used for determining the species: "Flora of the Lower Volga region" [2, 3], "Conspectus florae Europae Orientalis" [4], "Flora of a central part of the European Russia" [5].

* Corresponding author: stsenator@yandex.ru

The nomenclature and volume of taxa are mainly standardized by "The Euro+Med PlantBase" [6]. Abbreviations of taxa authors are given according to "International Plant Name Index" [7].

The geographical coordinates are given in WGS 84 system when quoting labels. The number of the square is indicated in brackets in Atlas Flora Europaea [8]. Collected herbarium specimens are stored in PVB; doublets are transferred to MW and NS.

## 3. Results

The studies revealed 565 species of vascular plants [9], 19 species of which are included in the Red Book of the Russian Federation [10], 28 species are in the Red Book of Volgograd oblast [1] and 2 species are in the List of European Red List of Vascular Plants [11].

Data on findings of rare vascular plants are presented below. Collection of vascular plants and the species inside are listed in order of the Latin alphabet. * – species listed in the Red Book of the Russian Federation [10].

**Asteraceae**

***Artemisia salsoloides*** Willd.*: 1) 49°02'54" N, 43°33'07" E, Kalachevsky dist., northern surroundings of the Evlampievsky farmstead, the upper parts of the Bolshaja Golubaja river, marl-chalk outcrops, 28.V.2018 [38ULV4]; 2) 49°10'59" N, 43°45'15" E, Ilovlinsky dist., Natural park "Donskoy", right bank of the Don river, marl-chalk outcrops, 29.V.2018 [38UMV2].

***Centaurea gerberi*** Steven: 1) 48°55'28" N, 43°47'40" E, Kalachevsky dist., surroundings of the Peskovatka village, sands on the left bank of the Don river, 25.V.2018 [38UMV2].

***Jacobaea schwetzowii*** (Korsh.) Tatanov et Vasjukov [*Senecio schwetzowii* Korsh.]: 1) 49°02'44" N, 43°32'47" E, Kalachevsky dist., northern surroundings of the Evlampievsky farmstead, the upper parts of the Bolshaja Golubaja river, forb-grass meadows in the floodplain, 27.V.2018 [38ULV4].

***Jurinea cretacea*** Bunge*: 1) 49°02'54" N, 43°33'07" E, Kalachevsky dist., northern surroundings of the Evlampievsky farmstead, the upper parts of the Bolshaja Golubaja river, marl-chalk outcrops, 28.V.2018 [38ULV4]; 2) 49°10'59" N, 43°45'15" E, Ilovlinsky dist., Natural park "Donskoy", right bank of the Don river, marl-chalk outcrops, 29.V.2018 [38UMV2].

***Jurinea ledebourii*** Bunge: 1) 49°00'39" N, 43°49'31" E, south of the Ilovlinsky dist., right bank of the Don river, feather-grass steppe with forbs, 29.V.2018 [38UMV2].

**Brassicaceae**

***Crambe tataria*** Sebeók: 1) 49°02'54" N, 43°33'07" E, Kalachevsky dist., northern surroundings of the Evlampievsky farmstead, the upper parts of the Bolshaja Golubaja river, marl-chalk outcrops, 28.V.2018 [38ULV4]. Listed in the European Red List of vascular plants, category LC [11].

***Lepidium coronopifolium*** Fisch. ex DC.: 1) 50°07'23" N, 46°19'25" E, Nikolaevsky dist., surroundings of the Libnekht village, to the south-west of the sheep barn "Kulturnaja", dry steppe dominated by *Festuca valesiaca*, *Stipa lessingiana*, and *Tanacetum achilleifolium*, 22.V.2018 [38UNA3].

***Lepidium meyeri*** Claus*: 1) 49°02'54" N, 43°33'07" E, Kalachevsky dist., northern surroundings of the Evlampievsky farmstead, the upper parts of the Bolshaja Golubaja river, marl-chalk outcrops, 28.V.2018 [38ULV4]; 2) 49°10'59" N, 43°45'15" E, Ilovlinsky dist., Natural park "Donskoy", right bank of the Don river, marl-chalk outcrops, 29.V.2018 [38UMV2].

***Matthiola fragrans*** (Fisch.) Bunge*: 1) 48°58'11.6" N, 43°38'02.7" E, north of the Kalachevsky district, left bank of the Bolshaja Golubaja river, chalk outcrops, 29.V.2018

[38UMV2]; 2) 49°10'59" N, 43°45'15" E, Ilovlinsky dist., Natural park "Donskoy", right bank of the Don river, marl-chalk outcrops, 30.V.2018 [38UMV2].

**Caryophyllaceae**

***Otites hellmannii*** (Claus) Stankov [*Silene hellmannii* Claus]*: 1) 49°02'54" N, 43°33'07" E, Kalachevsky dist., northern surroundings of the Evlampievsky farmstead, the upper parts of the Bolshaja Golubaja river, marl-chalk outcrops, 28.V.2018 [38ULV4]; 2) 49°10'59" N, 43°45'15" E, Ilovlinsky dist., Natural park "Donskoy", right bank of the Don river, marl-chalk outcrops, 30.V.2018 [38UMV2].

**Colchicaceae**

***Colchicum laetum*** Steven*: 1) 49°02'54" N, 43°33'07" E, Kalachevsky dist., northern surroundings of the Evlampievsky farmstead, the upper parts of the Bolshaja Golubaja river, feather-grass steppe with forbs, 28.V.2018 [38ULV4].

**Cupressaceae**

***Juniperus sabina*** L.: 1) 49°11'24.5"N, 43°36'04.4" E, Ilovlinsky dist., surroundings of the Sirotinskaja village, chalk outcrops, 30.V.2018 [38ULV4]; 2) 49°27'57" N, 43°41'14" E, Ilovlinsky dist., surroundings of the Viltov village, Archedinsko-Donskie sands, 30.V.2018 [38UMV1]; 3) 49°30'11" N, 43°38'46" E, Frolovsky dist., surroundings of the Zimovsky village, Archedinsko-Donskie sands, 31.V.2018 [38UMV1].

**Fabaceae**

***Genista tanaitica*** P. A. Smirn.*: 1) 49°02'54" N, 43°33'07" E, Kalachevsky dist., northern surroundings of the Evlampievsky farmstead, the upper parts of the Bolshaja Golubaja river, marl-chalk outcrops, 28.V.2018 [38ULV4]; 2) 49°10'59" N, 43°45'15" E, Ilovlinsky dist., Natural park "Donskoy", right bank of the Don river, marl-chalk outcrops, 29–30.V.2018 [38UMV2].

***Hedysarum cretaceum*** DC.*: 1) 48°58'11.6" N, 43°38'02.7" E, north of Kalachevsky dist., left bank of the Bolshaja Golubaja river, chalk outcrops, 29.V.2018 [38UMV2]; 2) 49°10'59" N, 43°45'15" E, Ilovlinsky dist., Natural park "Donskoy", right bank of the Don river, marl-chalk outcrops, 30.V.2018 [38UMV2].

***Hedysarum smirnovii*** Knjasev: 1) 49°02'54" N, 43°33'07" E, Kalachevsky dist., northern surroundings of the Evlampievsky farmstead, the upper parts of Bolshaja Golubaja river, marl-chalk outcrops, 28.V.2018 [38ULV4]; 2) 49°10'59" N, 43°45'15" E, Ilovlinsky dist., Natural park "Donskoy", right bank of the Don river, marl-chalk outcrops, 29.V.2018 [38UMV2]; 3) 49°11'24.5" N, 43°36'04.4" E, Ilovlinsky dist., surroundings of the Sirotinskaja village, chalk outcrops, 30.V.2018 [38ULV4].

Note. Hybridogenic species derived from hybridization *H. biebersteinii* Chrtková × *H. grandiflorum* Pall. It is recommended for inclusion in the new edition of the Red Book of Volgograd oblast.

**Iridaceae**

***Iris pumila*** L.*: 1) 48°50'31" N, 43°51'08" E, Gorodishchensky dist., between the Peskovatka and Dmitrievka villages, forb and feather-grass steppe in the Rubezhnaja balka, 27.V.2018 [38UMV2]; 2) 49°02'44" N, 43°32'47" E, Kalachevsky dist., northern surroundings of the Evlampievsky farmstead, the upper parts of the Bolshaja Golubaja river, feather-grass steppe with forbs, 27.V.2018 [38ULV4].

**Lamiaceae**

***Hyssopus cretaceus*** Dub.*: 1) 49°02'54" N, 43°33'07" E, Kalachevsky dist., northern surroundings of the Evlampievsky farmstead, the upper parts of the Bolshaja Golubaja river, marl-chalk outcrops, 28.V.2018 [38ULV4].

**Liliaceae**

***Fritillaria ruthenica*** Wikstr.*: 1) 50°07'23" N, 46°19'25" E, Nikolaevsky dist., surroundings of the Libnekht village, to the south-west of the sheep barn "Kulturnaja", dry steppe dominated by *Festuca valesiaca*, *Stipa lessingiana*, and *Tanacetum achilleifolium* on moistened microdepressions along the Rybnaja Soljanka river, 22.V.2018 [38UNA3]; 2) 49°00'39" N, 43°49'31" E, south of Ilovlinsky dist., right bank of the Don river, oak and maple bairak forest, 29.V.2018 [38UMV2].

***Tulipa schrenkii*** Regel [*T. gesneriana* auct. non L.]*: 1) 50°07'23" N, 46°19'25" E, Nikolaevsky dist., surroundings of the Libnekht village, to the south-west of the sheep barn "Kulturnaja", dry steppe dominated by *Festuca valesiaca*, *Stipa lessingiana*, and *Tanacetum achilleifolium*, 22.V.2018; 2) 49°25'17" N, 46°01'00" E, Pallasovsky dist., north-west surroundings of the Bulukhta lake, Yudina stow, herb-bunchgrass steppe with *Tanacetum achilleifolium* and forb-grass steppe with *Artemisia lerchiana*, 23.V.2018 [38UNV3]; 3) 48°41'58" N, 45°31'56" E, Leninsky dist., 3 km to the north of the Kolobovka village, desert steppe, 24.V.2018 [38UNU1].

**Poaceae**

***Koeleria sclerophylla*** P. A. Smirn.*: 1) 49°02'54" N, 43°33'07" E, Kalachevsky dist., northern surroundings of the Evlampievsky farmstead, the upper parts of Bolshaja Golubaja river, marl-chalk outcrops, 28.V.2018 [38ULV4].

***Stipa cretacea*** P. A. Smirn.: 1) 49°02'54" N, 43°33'07" E, Kalachevsky dist., northern surroundings of the Evlampievsky farmstead, the upper parts of the Bolshaja Golubaja river, marl-chalk outcrops, 28.V.2018 [38ULV4]; 2) 49°10'59" N, 43°45'15" E, Ilovlinsky dist., Natural park "Donskoy", right bank of the Don river, marl-chalk outcrops, 29.V.2018 [38UMV2]; 3) 49°11'24.5" N, 43°36'04.4" E, Ilovlinsky dist., surroundings of the Sirotinskaja village, chalk outcrops, 30.V.2018 [38ULV4].

***Stipa dasyphylla*** (Czern. ex Lindem.) Trautv.*: 1) 49°00'39" N, 43°49'31" E, south of the Ilovlinsky dist., right bank of the Don river, feather-grass steppe with forbs, 29.V.2018 [38UMV2].

***Stipa pulcherrima*** K. Koch*: 1) 49°02'54" N, 43°33'07" E, Kalachevsky dist., northern surroundings of the Evlampievsky farmstead, the upper parts of the Bolshaja Golubaja river, feather-grass steppe with forbs, 28.V.2018 [38ULV4]; 2) 49°00'39" N, 43°49'31"E, south of the Ilovlinsky dist., right bank of the Don river, feather-grass steppe with forbs, 29.V.2018 [38UMV2]; 3) 49°10'59" N, 43°45'15" E, Ilovlinsky dist., Natural park "Donskoy", right bank of the Don river, forb-grass communities with *Artemisia lerchiana*, 29.V.2018 [38UMV2].

***Stipa zalesskii*** Wilensky ex P. A. Smirn.*: 1) 50°07'23" N, 46°19'25" E, Nikolaevsky dist., surroundings of the Libnekht village, to the south-west of the sheep barn "Kulturnaja", dry steppe dominated by *Festuca valesiaca*, *Stipa lessingiana*, and *Tanacetum achilleifolium*, 22.V.2018 [38UNA3]; 2) 49°02'54" N, 43°33'07" E, Kalachevsky dist., northern surroundings of the Evlampievsky farmstead, the upper parts of Bolshaja Golubaja river, feather-grass steppe with forbs, 28.V.2018 [38ULV4]. Listed in the European Red List of vascular plants, category DD [11].

**Ranunculaceae**

***Delphinium puniceum*** Pall.*: 1) 49°02'44" N, 43°32'47" E, Kalachevsky dist., northern surroundings of the Evlampievsky farmstead, the upper parts of the Bolshaja Golubaja river, feather-grass steppe with forbs, 27.V.2018 [38ULV4].

***Pulsatilla pratensis*** Mill.*: 1) 49°30'11" N, 43°38'46" E, Frolovsky dist., surroundings of the Zimovsky village, Archedinsko-Donskie sands, 31.V.2018 [38UMV1].

**Rubiaceae**

***Asperula tephrocarpa*** Popov et Chrshan.: 1) 49°02'54" N, 43°33'07" E, Kalachevsky dist., northern surroundings of the Evlampievsky farmstead, the upper parts of Bolshaja Golubaja river, marl-chalk outcrops, 28.V.2018 [38ULV4].

**Scrophularaceae**

***Scrophularia cretacea*** Fisch.*: 1) 49°02'54" N, 43°33'07" E, Kalachevsky dist., northern surroundings of the Evlampievsky farmstead, the upper parts of the Bolshaja Golubaja river, marl-chalk outcrops, 28.V.2018 [38ULV4]; 2) 49°10'59" N, 43°45'15" E, Ilovlinsky dist., Natural park "Donskoy", right bank of the Don river, marl-chalk outcrops, 29.V.2018 [38UMV2].

## 4. Discussion

As a result of the research, 13 localities of 28 species of vascular plants, listed in the Red Book of Volgograd oblast [1] were identified, 19 species of which are included in the Red Book of the Russian Federation [10] and 2 species are in the European Red List of vascular plants [11]. Most of the detected species is confined in its distribution to the chalk and marl-chalk outcrops. Each location is linked to the grid of Atlas Flora Europaea.

The obtained materials on the distribution of rare and threatened species of vascular plants are the important data for long-term monitoring, as well as for environmental measures at the regional level.

## 5. Acknowledgements

The work was carried out in the framework of the state task of Institute of Ecology of the Volga river basin of the RAS, theme № 0128-2014-0002.

## References

1. *The Red Book of Volgograd oblast. Plants and other organisms*, **2** (Volgograd, Publisher-Printer, 2017)
2. *Flora of the Lower Volga region*, **1** (Moscow, KMK Sci. Publ., 2006)
3. *Flora of the Lower Volga region*, **2** (Moscow, KMK Sci. Publ., 2018)
4. *Conspectus florae Europae Orientalis*, **1** (Moscow; Saint-Petersbourg, 2012)
5. P.F. Maevskij, *Flora of a central part of the European Russia*, 11th ed. (Moscow, 2014)
6. Euro+Med (2006): Euro+Med PlantBase - the information resource for Euro-Mediterranean plant diversity. Published on the Internet http://ww2.bgbm.org/EuroPlusMed/ [accessed 18 July 2019].
7. The International Plant Names Index (2019). Published on the Internet http://www.ipni.org [accessed 18 July 2019]
8. Grids and mapping – Atlas florae Europaeae. https://www.luomus.fi/ [accessed 18 July 2019]
9. S.A. Senator, V.M. Vasjukov, E.G. Zibzeev, A.Y. Korolyuk, S.V. Saksonov, Phytodiversity of Eastern Europe, **XII**, 4 (2018)
10. *The Red Book of the Russian Federation (plants and fungi)* (Moscow, KMK Sci. Press, 2008
11. M. Bilz, S.P. Kell, N. Maxted, R.V. Lansdown, *European Red List of Vascular Plants* (Luxembourg, Publications Office of the European Union, 2011)

# To the issue of third edition of the Red List of the Kemerovo region

*Svetlana* Sheremetova* and *Irina* Khrustaleva

Department "Kuzbass Botanical Garden", Federal State Budget Scientific Institution «The Federal Research Center of Coal and Coal Chemistry of Siberian Branch of the Russian Academy of Sciences», 650065, Leningradsky av., 10, Kemerovo, Russia

**Abstract.** Recent systematic floristic researches cause the necessity of changes and amendments in the lists of rare and disappearing plants, included in the Red Book of the Kemerovo region of first (2000) and second (2012) edition. New recommendations for third edition are proposed. 142 species will be included into third edition of the Red Book of the Kemerovo region 32 more species will be added for the first time, 18 species will be removed. Only 46 species will keep both the presence and statuses (categories) of a rarity; 34 species, presented in all three lists, will change their category (status).

Botanical researches in Kuzbass began in the first half of the 18th century; during 300 years history they had both periods of rising and periods of recession. A.V. Kuminova in monograph "Vegetation of the Kemerovo region" [1] noted that by the middle of the 20th century the Kemerovo region was one of the most popular for investigations regions of Siberia. I.M. Krasnoborov [2] wrote that in the second half of XX century the interest in Kemerovo region botanical researches was lost. Nevertheless, by the beginning of the 21st century collected materials and capacity of scientific community allowed to prepare for publication the Red Book of the Kemerovo Region [3] (in 2000), and the «Plant identification guide» (in 2001) [4]. I.M. Krasnoborov [2] notes that work on writing «Plant identification guide» was very difficult, as there were very few floristic publications concerning some groups of plants of the region, and the herbarium collected across Kuzbass was not full. Moreover, the collection was storaged in different Herbariums of the country.

Regular floristic investigations on the territory of the region, carried out by researchers of the Kuzbass botanical garden [5], began in 2002. The most complete Herbarium (KUZ) of the Kemerovo region with more than 16000 samples of vascular plants [6] of Kuzbass was created.

New information about known location number, new locations, and new species [7-9] allowed to add amendments to the list of rare plants in second edition of the Red Book of the Kemerovo region (2012) [10]. 29 species were removed from the list of rare and disappearing plants of the Kemerovo region [3], the status was changed for 29 species, 27 species were included in the list for the first time (14 of them were new to the region). The

* Corresponding author: ssheremetova@rambler.ru

section of resource plants was excluded. 9 new families represented by the following species: Aristolochiaceae (*Asarum europaeum* L.), Asparagaceae (*Asparagus pallasii* Miscz.), Convolvulaceae (*Convolvulus chinensis* Ker-Gawl.), Fumariaceae (*Corydalis solida* subsp. *subremota* (M. Popov) Peschkova), Linaceae (*Linum perenne* L., and in the third the edition – *Linum altaicum* Ledeb. ex Juz.), Polygalaceae (*Polygala tenuifolia* Willd.), Saxifragaceae (*Mitella nuda* L.), Cupressaceae (*Juniperus communis* L.), Athyriaceae (*Cystopteris altajensis* Gureeva). The species list of some families was completely changed: for example, *Allium ramosum* L. and *Allium vodopjanovae* Friesen were added instead of *Allium rubens* Schrader ex Willd. and *Allium clathratum* Ledeb. 4 families with the following species were removed: Apocynaceae (*Apocynum lancifolium* Russanov), Gentianaceae (*Gentiana septemfida* Pall.), Limoniaceae (*Limonium gmelinii* (Willd.) O. Kuntze), Huperziaceae (*Huperzia selago* Bernh. ex Schrank et Mart).

As results of work of researchers from Kuzbass Botanical Garden (S.A. Sheremetova, A.N Kupriyanov.), Central Siderian Botanical Garden (N.N. Lashchinsky), Tomsk State University (A.L. Ebel), Kemerovo State University (A.V. Klimov) recommendations to the third edition of the Red List of the Kemerovo region were created.

For inclusion in the list of rare and disappearing plants of the Kemerovo region 32 species are recommended (the recommended status is given in brackets): *Astragalus vaginatus* Pall. (1), *Astragalus austriacus* Jacq. (2), *Stipa lessingiana* Trin. et Rupr. (2), *S. praecapillata* Alechin (2), *Agropyron kazachstanikum* (Tzvel.) Peschkova (3), *Astragalus follicularis* Pall. (3), *Thellumgiella salsuginea* (Pallas) O.E. Schulz (3), *Blysmus rufus* (Hudson) Link (3), *Arctous erythrocarpa* Small (3), *Sajanella monstrosa* (Willd. ex Sprengel) Sojak (3), *Rhodiola quadrifida* (Pall.) Fisch.et Mey. (2)., *Linum altaicum* Ledeb. ex Juz. (1), *Carex alba* Scop. (3), *C. buxbaumii* Wahlenb. (2), *C. serotina* Merat (2), *C. sedakowii* S.A. Meuyeg of ex Meinsh. (3), *Potentilla erecta* (L.) Raeuschel (2), *Anemonidium dichotomum* (L.) Holub (3), *Cypripedium venthcosum* Sw. (3), *Gymnadenia conopsea* (L.) R. Br. (3), *Herminium monorchis* (L.) R. Br. (3), *Botrychium virginianum* (L.) Sw. (3). With a part of new types such families as also are included in the list: Menyanthaceae (*Nymphoides peltata* (S.G. Gmel.) Kuntze (3)), Salicaceae (*Populus alba* L. (3)), Chenopodiaceae (*Atriplex fera* (L.) Bunge (1), *Krascheninnikovia ceratoides* (L.) Gueldenst. (1)), Portulacaceae (*Claytonia Joanneana* Roem. et Schult. (3)), Lentibulariaceae (*Utricularia minor* L. (2)), Thelypteridaceae (*Oreopteris limbosperma* (All.) Holub. (3)), Woodsiaceae (*Woodsia pinnatifida* (Fomin) Schmakov (3), *W. heterophylla* (Turcz. ex Fomin) Schmakov (2) *W. calcarea* (Fomin) Schmakov (3)).

Change of category (status) of rarity is recommended to 15 species: *Lupinaster eximius* (Steph. ex Serg.) C. Presl. (from 2 → to 1), *Rhododendron aureum* Georgi (3 → 1),*Cerastium maximum* L. (3 → 2), *Geranium robertianum* L. (2 → 3), *Bupleurum bicaule* Helm (2 → 1), *Trinia ramosissima* Ledeb. (2 → 1), *Erysimum flavum* subsp. *altaicum* (C.A. Meyer) Polozhij (2 → 3), *Fritillaria meleagroides* Patrin ex Schult.et Schult. (2 → 3), *Tulipa patens* Agardh ex Shult.et Shult. (1 → 2), *Scrophularia umbrosa* Dumort.(1 → 3), *Dactylorhiza russowii* (Klinge) Holub (2 → 3), *Viola dissecta* Ledeb.(2 → 3), *Ephedra monosperma* C. A. Mey. (2 → 3), *Ophioglossum vulgatum* L. (1 → 3), *Hedysarum turczaninovii* Peschkova (3 → 2).

We recommend excluding from third edition of 18 species of vascular plants.: *Artemisia pontica* L, *Artemisia santolinifolia* Turcz. ex Bess., *Corydalis solida* subsp. *subremota* (M. Popov) Peschkova, *Dactylorhiza fuchsii* (Druce) Soó, *D. longifolia* (L. Neum.)Aver., *Elatine triandra* Schkuhr, *Hordeum roshevitzii* Bowden, *Lindernia procumbens* (Krock.) Borb, *Myosotis jenissejrnsis* O. D. Nikiforova, *Paeonia hybrida* Pall. *Primula patens* (Turcz.)E. Busch (*P. sieboldii* E. Morren), *Pyrola media* Sw., *Rhododendron Ledebourii* Pojark., *Spiraea crenata* L., *Trapa natans* L., *Viola incisa* Turcz. *Scirpus validus* Vahl, *Lathyrus krylovii* Serg. Some of them are not confirmed by herbarium collection; some species were found in good condition of large local populations, some taxa are still dubious in the modern systematics. For example, *Corydalis solida* subsp. *subremota* was included into the Red Book (2012) with rare indication of locations near village Ankimo-Annenka according to the herbarium of Kemerovo State University. Recent researches showed no finding near this village, but this species is widespread in Tyazhin river basin: state of its populations allow to recommend it for removal from list. Some families in the Red Book of the Kemerovo region (2012) were presented by one species, and in the third edition, after the accepted changes five families will be removed: Elatinaceae, Paeoniaceae, Primulaceae, Trapaceae, Fumariaceae.

Despite the accepted changes and new recommendions, 46 species keep not only the positions in the list of rare and disappearing plants, but also the statuses (categories) of rarity. 34 species presented in all three lists of the Red Books changed categories, the status of a rarity was increased for 11 of them.

Thus, the Red Book of the Kemerovo region included 128 species of vascular plants in first and second editions, 142 species were recommended for third edition. In third edition 56 % of the species is supposed to remain on the list, but only 32 % keep the category and the rarely status. We consider that the accumulation of information about rare plants species allows to reduce the subjectivity of rare species and rare species status.

## Acknowledgment

Work is performed within the State task № 0352-2019-0015, with involvement of funds of USI «Herbarium of the Kuzbass botanical garden» (KUZ)

## References

1. A.V. Kuminova. T*he Vegetation of the Kemerovo region* (Novosibirsk, 1950)
2. I.M. Krasnoborov, Botanical research in Siberia and Kazakhstan, **12** (2006)
3. *Red Book of the Kemerovo region* (Kemerovskoe kniznoe izdatelstvo, Kemerovo, 2000)
4. *Plant identification guide of the Kemerovo region* (Publishing House SB RAS, Novosibirsk, 2001)
5. S.A. Scheremetova, *Flora of the Tom River Basin: composition, structure, transformation, spatial organization* (Tomsk, 2016)
6. I.A. Khrustaleva, Botanical collections - a national treasure of Russia, 285 (2015)
7. T.E. Buko, S.A. Scheremetova, A.L. Ebel, Botanical research in Siberia and Kazakhstan, **15** (2009)
8. S.A. Scheremetova, A.L. Ebel, T.E. Buko, Turczaninowia, **14(1)** (2011)
9. A.L. Ebel, Turczaninowia, **5(2)** (2002)

1. *Red Book of the Kemerovo region* (Azya print, Kemerovo, 2012)

# Dynamics of the adventive component in the flora of Barnaul

*Marina* Silanteva*, and *Tatyana* Terekhina

Altai State University, Botany Department, 61 Lenin Str., Barnaul, 656049, Russia

**Abstract.** The article considers the results of the floristic study as well as the analysis of the archival records and literature data in order to reveal the dynamics and the specific features of the adventive element formation in the flora of Barnaul throughout its history. It provides data on the time of some species introduction into the city area.

## Introduction

It is necessary to turn to the history of the Barnaul city in order to understand the specific aspects of the adventive element formation on its territory. Barnaul was founded (according to one theory) in 1730 by the Ural manufacturer A. Demidov. The construction of the Barnaul silver-smelting plant in 1739–1744 was one of the steps in the industrial mining development of the Altai territory, which did not have cities and large settlements as well as developed agricultural production before that. Intensive colonization of the Altai territory by the Russian people began only in the 1720s. It was closely associated with the discovery of rich copper and silver ore deposits in the northwestern foothills (the city of Zmeinogorsk and its outskirts). It was in 1749, when the executive office of the entire newly created mountain district was transferred to Barnaul and the city began to develop as an industrial and administrative center. A technical library comprising 7 thousand volumes was opened in 1764. In 1771, Barnaul gained the status of a mountain city where the first school of mines in Siberia was established. Archival records from 1786 mention the Barnaul "theater house".

Initially, the area of the city was limited by the Ob River in the east and southeast, and by the winding forest in the southwest and west, so that the city grew northwards. The city was developing in the Barnaulka River valley, close to its mouth (the confluence of the Barnaulka and the Ob Rivers), claiming flood meadows, dry meadows of terraces and birch as well as pine forests. In terms of natural conditions, the territory of the city is the right bank forest-steppe.

## Results

In 1779, a corresponding member of the St. Petersburg Academy of Sciences E. Patren compiled the first inventory of plants in the vicinity of Barnaul *"Flora barnaulensis"*. This

* Corresponding author: msilan@mail.ru

catalog specifies 244 plant species [1]. The publications of the XVIII century preserved the information on what the inhabitants of the city cultivated. Thus, in 1796, I. Sivers, a corresponding member of the St. Petersburg Academy of Sciences, provided the data on the cultivation of strawberries, cucumbers, melons and watermelons in Barnaul in "Letters from Barnaul" published in the "Works of the Free Economic Society".

The introduction of adventive species was also due to the pharmacy gardens. In 1800, S. Shangin set out a small botanical garden with a greenhouse for thermophytes in the place of the pharmacy garden at the Barnaul Central Hospital. Up to 400 species of plants were grown in the botanical garden; there was a plantation of rhubarb [2, 3].

At the beginning of the XIX century, the city became a major cultural, scientific and engineering center of Russia. In 1827, the first printing office and the museum of regional studies, one of the first in Siberia, were established. Apparently, the arriving Cossacks and mountain officers, graduates of St. Petersburg Mining Institute, brought various planting material from different places of Russia. In the published list of the Altai Mountains plants, probably determined by Fisher himself [1], I. Brykov [4], a future famous doctor and writer, indicated, along with the cultivated *Convallaria majalis, Borago officinalis, Coryllus avellana, Aquilegia vulgaris*, etc., a number of weed adventive species introduced together with the cultivated ones: *Agrostemma githago, Echium vulgare, Betonica officinalis, Malva crispa, Raphanus raphanistrum, Lamium purpureum, Lepidium ruderale, Brassica campestris, Sinapis arvensis, Sonchus oleraceus, Fumaria officinalis, Centaurea cyanus, C. jacea, Matricaria chamomilla, Anthemis tictoria, Asperugo procumbens*, and others.

In 1835, more than 9 thousand people lived in Barnaul. In 1864, S. I. Gulyaev published a note on the development of gardening and horticulture in Barnaul. Mentioning that vegetable gardening serves merely the needs of the townspeople themselves, he reported that they sold cucumbers, melons, watermelons, potatoes and cabbage. The list of vegetable crops grown by the people in Barnaul was significant. It was specified that "gardening in its direct meaning did not exist in Barnaul", "and private gardens looked like small gardens in St. Petersburg on the Vyborg side, in Koltovskaya, Kolomna and Gavan". In gardens and vegetable gardens, one could see the birch, bird cherry, linden (from the Kuznetsk district), pine, cedar, mountain ash, buckthorn, guilder rose, honeysuckle, meadowsweet, shadbush, coloneaster, Kalmyk peaches (almonds), sea buckthorn, and hawthorn. One cannot but mention raspberries, currants, gooseberries among berry-producing shrubs and the Siberian apple tree among fruit trees [5]. At the same time, a future famous botanist A. N. Krasnov specified *Malva borealis, Lycopsis arvensis, Elsholtia ciliata, Urtica cannabina* as adventive weeds introduced to Barnaul by people [6].

By the 1870s, the city's population reached 13-14 thousand people. The decline of the mining industry and the shutdown of the Barnaul plant in 1893 led to a shift in the role of the city. It became a major trading center of a developing agricultural region.

In 1885–1890, the oldest amateur gardens of Barnaul were set out. Both seedlings of European fruit trees varieties and forms of Siberian apple trees were grown there. Wide-scale planting of gardens in the region began in 1894–1902, when the district headquarters began to lease land plots in the outskirts of the city for long-term periods of 24 years [7]. A lot of dachas (summerhouses) and zayimkas (small isolated settlements on uncultivated land or hunters' cabins) appeared in the suburbs of Barnaul. Gardeners of that time introduced such species as pedunculate (common) oak, European white elm, balsam poplar, Tartarian and Norway maple, small-leaved and Siberian linden, Siberian apple and its hybrids, shadbush, sloe, common lilac, mock orange and many others [8].

After the Civil war of 1917-1919 and the catastrophic fire of 1917, the city in fact, turned into the center of the region's agro-industrial complex, and vegetable gardens became an integral part of the city landscape.

In 1930, V. I. Vereschagin became the second researcher in the history of Barnaul to create its flora. However generally, we accept his list as the first "Flora of Barnaul", since Patren's list, preserved in the archives, was not in common scientific use. "The List of Plants in the Vicinity of Barnaul" by V. I. Vereschagin [9] was compiled on the basis of the materials collected on the left bank of the Ob in 1900-1929 and on the right bank in the flood meadows along the Talaya and the Losikha Rivers. The list contains 743 species.

In the post-war period, gardening quickly developed in Altai. The process became especially intense since the 1960s. M. A. Lisavenko Siberian Research Institute of Horticulture (SRIofH) made a particularly outstanding contribution to the enrichment of the city's flora with decorative trees and shrubs. The basis of the woody plants assortment was presented by 33 species of trees and more than 60 species of shrubs [7]. A number of introduced species (European white elm, common oak, Pallas apple tree, small-leaved linden, green ash, maple ash, etc.) were able to regenerate naturally even in the conditions of the forest-steppe and now constitute an adventive element of the city's flora.

In the 80-90s of the last century, there was a steady rise in private landholding (gardens and summerhouses, estates). The interest of the population in landscaping with the use of phytodesign techniques and the use of exotic planting and seed material increased. Spontaneous introduction is currently significant. In the vicinity of the city and in its streets, one can see *Thladiantha dubia*, cultivars of *Aquilegia vulgaris, Ipomea purpurea, Xanthoxalis stricta*, etc. at landfill sites.

Currently, the population of the city is about 700,000 people. In the city area, the vegetation cover is represented mainly by artificial plantations – alleys, public gardens, boulevards, lawns and parks. The outermost northeastern part of the Barnaul pine forest formed by the common pine is situated in the upland part of the city along the Barnaulka River on the sandy ridges. The area of Nauchnyi Gorodok is characterized by aspen-birch outliers and steppe plots. The city is surrounded by horticultural and agricultural lands.

The study of the city flora in the 1990s by the Botany Department team allowed identifying only 578 species of higher vascular plants within the precincts of the city. In our opinion, such an outcome is the result of dense urban development, a decrease in the share of natural phytocenoses and an increase in the share of anthropogenically transformed ones, unification of landscaping, and a reduction in the number of private residences. Native species of natural habitats that cannot tolerate an increase in anthropogenic load keep falling out. Alien species and escapees from cultivation amounting to 90 species (14.5%) dominated among the newly introduced plants. All these changes indicate an increase in the degree of the flora anthropogenic transformation. By the beginning of the 21st century, 200 adventive plants belonging to 48 families and 143 genera were identified (34.3% of all flora species).

In the 30s-50s of the XX century, 55 adventive plant species of 46 genera and 17 families grew in the city. Most of the families were represented by 1-3 species (*Cuscutaceae, Fumariaceae, Violaceae, Fumariaceae, Rubiaceae, Amaranthaceae, Polygonaceae, Malvaceae, Fabaceae, Boraginaceae, Lamiaceae, Solanaceae*). From 4 to 14 species belonged to the families of *Chenopodiaceae, Caryophyllaceae, Poaceae, Asteraceae, Brassicaceae*.

In the 1990s, new adventive plants of the *Ulmaceae, Cannabaceae, Aceraceae, Balsaminaceae, Elaeagnaceae, Onagraceae, Apiaceae, Cucurbitaceae, Gentianaceae, Papaveraceae* families appeared. They were represented by one species each. Previously identified adventive species also retained their representation in general. In total, the representatives of 28 families and 68 genera of adventive plants grew on the territory of the city during this period [10].

By the beginning of the XXI century, there was a sharp rise in the number of adventive species in the city's flora. New species representing the families (1 species per a family) of

*Commelinaceae, Salicaceae, Plumbaginaceae, Urticaceae, Portulacaceae, Fagaceae, Juglandaceae, Vitaceae, Tiliaceae, Hydrangeaceae, Rutaceae, Scrophulariaceae, Plantaginaceae, Oleacae, Campanulaceae, Euphorbiaceae, Ranunculaceae, Grossulariaceae, Rosaceae* were identified. A part of the adventive plant species transformed into the category of invasive ones and significantly increased their abundance. Thus, *Hordeum jubatum* L. has been found in Barnaul since 1977 and has become a common species on lawns and wastelands. At the end of the last century, *Echinocystis lobata* (Michx.) Torr. et Gray was encountered accidentally in shady ravines, along river banks, at landfill sites near the Gonba village. Currently it is common along the Ob floodplain near horticultural communities. Over a hundred-year period of use, it has become a common sight in the city, propagating by seeds. *Portulaca oleracea* L. was rarely found in vegetable gardens, on the lawns, flowerbeds, and along the roadsides. Currently, in 10 years' time, it has sharply increased its abundance, forming a continuous cover along the roads throughout the city. Birds actively spread *Amelanchier spicata* (Lam.) C. Koch from old gardens. *Acer negundo* L. forms undergrowth all over the forest, as well as in the city parks and in horticultural communities on uncultivated plots. Birds contributed to its active settlement. *Pastinaca sylvestris* Mill. is an ordinary weed in vegetable gardens, and it forms continuous thickets along the roads. Recently, *Cyclachaena xanthifolia* (Nutt.) Fresen has been identified in Barnaul. In the south of the region, it occupied wastelands and roadsides; in Barnaul, it is very active in all synanthropic habitats, forming thickets. *Solidago canadensis* L. has long been grown in gardens and escapes cultivation occupying abandoned plots and roadsides.

## Conclusion

The formation of the adventive species complex in the flora of Barnaul has its own specificity. It is determined by the types of nature management, transformed in different historical periods and leading to the creation of new habitats: ore processing; creation of hydraulic systems (dams, ponds); deforestation and afforestation; gardening, horticulture, floriculture; formation of a transport network; expansion of the built-up area and recreational territories. These new habitats have become a springboard for the introduction of a number of species. About 15 species of the identified 200 adventive plants species in Barnaul have become invasive in recent years.

## References

1. I. Borodin, *Collectors and collections of the Siberian flora* (St. Petersburg, 1908)
2. G. I. Kurnykina, I. P. Shangin, *Daily notes to the executive office of Kolyvano-Voskresenskiy mountain authorities on the trip to Kirgiz-Kaysatskaya steppe* (Barnaul, Az Buka, 2003)
3. N. A. Kambalov, Alt., **4 (39)**, 98 (1966)
4. I. Brykov, Uk. otkr. po fiz., khim., yest. ist. i tekhnol., izd. N. Scheglovym, **7/1**, (1830)
5. S. Gulyaev, VROS in StPbg, **1**, 15 (1864)
6. A. N. Krasnov, Botan. zap., **1**, 181 (St. Petersburg, 1886).
7. Z. I. Luchnik, Byull. GBS, **142**, 12 (1986)
8. Z. I. Luchnik, *Introduction of trees and shrubs in Altai Region* (1970)
9. V. I. Vereschagin, *The list of plants in the vicinity of Barnaul. Alt. collection*, **12** (1930)
10. T. A. Terekhina, *Anthropogenic phytosystems* (2000)

# Influence of ground fires on the grass cover of birch forests of the Eltsovsky discrit of the Altai region

*Galina* Sokolova*

Altai State University, Lenina str., 101, Barnaul, 656041, Russia

**Abstract**. The article contains the results of research of influence of surface fire on the vegetation of birch forests in the Eltsovsky district of the Altai region. Changes in species composition and structure of birch stands after fires on the slopes of different exposures are considered. Groups of species were identified depending on the reaction to the action of ground fires.

The study area covers the belt of birch forests and steppe meadows. It is characterized by hilly-rolling relief, strongly dissected by a network of valleys and beams. Birch forests, sometimes with an aspen admixture, on dark gray forest soils are the main type of vegetation. Steppe meadows on podzolized chernozem are developed on the watersheds, and on the slopes of the southern exposition there are areas of meadow steppes on leached chernozem.

Currently, the forest area has been greatly reduced due to clearings and the allocation of vacant land for arable land. The remaining birch forests are mainly located on the slopes of valleys and beams. For the release of agricultural land from plant residues often use their direct combustion. Agricultural burning spreads beyond arable land, leading to outbreak fires in forest ecosystems [1–3].

The aim of our work was to study the influence of low-level fires on the grass cover of birch forests in the Eltsovsky district of the Altai region. We carried out a reconnaissance survey of the area with the allocation of 10 control sites, reflecting the nature of the distribution and structure of birch forests, and 10 sites where forest ecosystems were exposed to low-level fires. Standard geobotanical releves of vegetation have been carried out at all sites [4]. Several plots, where the influence of ground fires on the grass cover and the dynamics of overgrowth of burning were being traced, were laid out on the slopes of different exposures.

There are two types of birch forests in the study area: birch forests with the predominance of meadow and meadow-steppe species growing on the southern slopes, and birch forests with large grass species typical of blackwood forests, and the northern exposure.

Such species as *Caragana arborescens* Lam., *Rosa majalis Herrm., R. acicularis* Lindl., *Spiraea media* F. Schmidt, etc. can be found in the undergrowth of birch forests. The first type of the grass cover of birch trees (whose height is 50–80 cm) is represented by the

* Corresponding author: sokolova-gg@mail.ru

following species: *Dactylis glomerata* L., *Calamagrostis arundinacea* (L.) Roth, *Trifolium lupinaster* L., *Bupleurum aureum* Fisch. ex Hoffm., *Vicia unijuga* A.Br., *Polygonatum odoratum* (Mill.) Druce, *Rubus saxatilis* L., *Crepis sibirica* L., *Pedicularis incinata* Steph., *Origanum vulgare* L., *Serratula coronata* L., *Phlomis tuberosa* L., *Pteridium aquilinum* (L.) Kuhn, etc.

The second type of the grass cover of birch forests with an average height of 120-150 cm is characterized by the presence of *Dactylis glomerata, Vicia sylvatica* L., *Crepis sibirica, Heracleum dissectum* Ledeb., *Pleurospermum urlense* Hoffm., *Cirsium heterophyllum* (L.) Hill, *Aconitum septentrionale* Koelle, *Euphorbia pilosa* L., *Pteridium aquilinum* (L.) Kuhn [5].

The analysis of the species composition and structure of the grass cover of the birch forests of the Eltsovsky district showed that the grass stand of birch forests is represented by 108 species of grassy plants belonging to 38 families and 87 genera. They are characterized by an uneven distribution over the slopes of different exposures. The greatest species diversity of birch forest cover is observed on the slopes of the southern exposures (43 species), the smallest number of species (18.0) is a characteristic of the slopes of the northern exposures. The number of species on the eastern and western slopes averages 28.

The total projective cover of birch stands on the slopes of the southern, eastern and western exposures averages 63.5 %, and on the slopes of the northern exposures – 33.7 %. The number of weed species ranges from 11.5 to 13.6 % of the total number of species. This indicator is slightly different for the slopes of different exposures. The share of rare and endangered species in the herbage on the slopes of all exposures averages 4.2%.

Such an indicator as the abundance of the species also varies depending on the exposure of the slope. *Dactylis glomerata, Pteridium aquilinum, Aegopodium podagraria* L. dominate on the slopes of the southern, eastern and western exposures. *Angelica decurrens* (Ledeb.) B. Fedtsch., *Aegopodium podagraria, Dactylis glomerata, Pteridium aquilinum* dominate on the slopes of the northern exposures.

Botanical analysis of the grass cover of birch forests has showed that forbs prevail on the slopes of all exposures. Grasses are more presented on the slopes of the eastern and southern exposures. The distribution of legumes and sedges goes on the slopes of all exposures in approximately the same way, which indicates good adaptive abilities of the species of these botanical groups.

Ecological analysis has revealed the prevalence of mesophytes on all slopes, their proportion ranging from 77.4 to 89.9 %. Group of mesohygrophytes falls out on the slopes of the southern exposure, since the hydrological conditions are not suitable for the representatives of this group. On the northern slopes the proportion of mesohygrophytes is maximum. Mesoxerophytes are present on the slopes of all exposures, but their share is higher on the western slopes.

The analysis of plant life forms revealed that perennial rhizomatous species prevail on the slopes of all exposures (96.7–97.8 %), the share of one- and two-year species is 2.2–3.3 %.

The study of the grassland of birch forests after fires showed that the species diversity of grass cover on the burns has decreased to 82 species belonging to 68 genera and 28 families. The tendency to reduce species diversity on the slopes of the western exposure is obvious. The greatest species diversity is typical of the slopes of the eastern and northern exposures, where the number of species has decreased only by 6 %, which indicates the best regenerative ability of the stand of birch forests in these conditions.

Fires in birch forests caused a qualitative change in the species composition of grassy vegetation, which means the loss of some plant species and the appearance of new species that had not been observed before the fire in the herbage.

The following types of plants – *Equisetum hyemale* L., *Sedum telephium* L., *Chamerion angustifolim* (L.) Holub, *Heracleum dissectum, Lamium album* L., *Sonchus arvensis* L., *Sonchus oleraceus* L., *Hemerocallis minor* Mill.- can be found in the birch forests on the slopes of southern exposure on the burned areas appear in. The slopes of northern exposition are represented by *Silene nutans* L., *Fragaria viridis* Duch., *Geum aleppicum* Jacq., *Chamerion angustifolium, Pedicularis incinata, Galeopsis bifida* Boenn., *Sonchus oleraceus.* Among the new species on the slopes of the eastern exposures are *Milium effusum* L., *Ranunculus repens* L., *Cannabis sativa* L., *Filipendula ulmaria* (L.) Maxim., *Agrimonia pilosa* Ledeb., *Epilobium montanum* L., *Chamerion angustifolium, Leucanthemum vulgare* Lam., *Artemisia vulgaris* L., *Sonchus oleraceus, Crepis sibirica, Crepis lyrata* (L.) Froel., *Sonchus arvensis.* The slopes of the western exposures are presented by such plants as *Calystegia sepium* (L.) R. Br., *Vicia craca* L., *Chamerion angustifolium, Anthriscus sylvestris* (L.) Hoffm., *Polemonium caeruleum* L., *Pedicularis incinata, Galeopsis bifida, Lamium album, Origanum vulgare, Artemisia vulgaris, Cacalia hastata* L. [6].

The total projective cover on the burns in birch forests has been reduced by 15 % compared to the control, and on the slopes of western exposures – by 27 %. The number of weed species on burns has increased by 5 % in comparison with the control and is 18 % of the total number of species. This tendency is typical of the slopes of the northern, southern and eastern exposures. As for the slopes of the western exposures the number of weed species does not differ significantly from the control.

The increase in the number of weed species on the burns in birch forests occurs due to the introduction of the following species – *Urtica dioica* L., *Sonhus oleraceus, Sonchus arvensis, Stellaria graminea* L., *Silene nutans, Oberna behen* (L.) Ikonn., *Agrimonia pilosa, Elytrigia repens* (L.) Nevski. On the burns, the projective cover and the occurrence of rare and endangered species have been reduced by 1.5 % compared with the control. The loss of such species as *Erythronium sibiricum* (Fisch. et May) Kryl., *Trollius asiatica* L., *Dactylorhiza baltica* (Klinge) Orlova, etc. is widely spread.

The burns of the first year are dominated by the following species: on the southern exposure slopes – *Dactylis glomerata, Pteridium aquilinum, Chamerion angustifolium, Aegopodium podagraria*; on the northern slopes – *Urtica dioica, Pteridium aquilinum, Chamerion angustifolium, Aegopodium podagraria;* on the slopes of the eastern exposure – *Urtica dioica, Aegopodium podagraria*.

The burns of the second year are mainly presented by: *Urtica dioica, Dactylis glomerata, Pteridium aquilinum, Chamerion angustifolium, Aegopodium podagraria* - on the southern exposure slopes; *Equisetum sylvaticum, Urtica dioica, Pteridium aquilinum, Chamerion angustifolium, Aegopodium podagraria* – on the northern slopes of exposure; *Artemisia vulgaris, Sonchus arvensis, Chamerion angustifolium* – on the eastern slopes; *Artemisia vulgaris, Aegopodium podagraria, Dactylis glomerata, Chamerion angustifolium, Pteridium aquilinum* – on the slopes of the western exposure.

The analysis of changes in the species composition and structure of the phytocenoses of the birch forests of the Eltsovsky region after the fire has revealed that the total number of burns in the burning areas has decreased on average by 6% in comparison with the control; total projective cover – by 15 %; the amount of weed species – by 5 %. The weed and meadow species of *Urtica dioica, Sonhus oleraceus, Sonchus arvensis, Stellaria graminea, Silene nutans, Oberna behen, Agrimonia pilosa, Elytrigia repens* are actively introduced into the grass stand.

Botanical analysis of the stand on the burns has revealed the following trends: an increase in the number of species and the total projective cover of cereals and a decrease in the total projective cover and the number of legumes. As for the grass the situation is ambiguous. The background of the general decrease in the projective cover and the number of burnt-out species are foolowed by an increase in the projective cover of the following species: *Aegopodium podagraria, Polygonatum odoratum, Crepis lyrata, Sonhus oleraceus, Artemisia vulgaris, Oberna behen, Glechoma hederaceae* L., *Dracocephalum ruyshiana* L., *Urtica dioica, Equisetum hyemale, Equisetum pratense* Ehrh., *Chamerion angustifolium, Polemonium caeruleum, Stellaria graminea, Silene nutans, Agrimonia pilosa, Elytrigia repens.*

Ecological analysis has showed that burns on the slopes of southern, northern and western exposures differ from the control by lower participation of mesophytes (by an average of 7 %) and mesoxerophytes (by an average of 2.4 %). There is a more significant decrease in the proportion of mesohygrophytes (by 9 %) on the slopes of northern exposures.

The two-year analysis of changes in the species composition of the grass cover of birch forests after fires has showed that not all species respond equally to this factor. According to the reaction of grassy plants to the action of agricultural fires, the following groups of species are distinguished:

– increasing numbers on burns – *Urtica dioica, Dactylis glomerata, Phleum pratense, Silene nutans, Aegopodium podagraria, Hemerocallis lilioasphodeus;*

– reducing the number on burns – *Trifolium lupinaster, Lathyrus gmelini* Fritsch., *Lathyrus pratensis, Thalictrum simplex* L., *Paris quadrifolia* L.*;*

– not changing the number on burns – *Origanum vulgare, Ptarmica impatiens* (L.) DC*;*

– appearing on burns – *Pedicularis incarnata, Hamerion angustifolium;*

– disappearing after fires – *Platanthera bifolia* (L.) Rich., *Dactylorhisa baltica.*

Thus, agricultural fires in the form of lowland fires have a significant impact on the species composition and grass stand of the birch forests of the Eltsovsky region, lead to a decrease in species and phytocenotic diversity, level the ecological structure of the phytocenoses, reduce the resumption of forest stand, and reduce the sustainability of forest ecosystems.

## References

1. E.N. Valendik, Siberian Journal of Ecology, 3 (1) (1996)
2. V.V. Furyaev, *The role of fires in the process of forest formation* (Novosibirsk: Science, 1996)
3. L.I. Sukhorukova, Yu.E. Belenets, V.A. Bochkov, N.Yu. Kobrin, Tr. St. Petersburg Institute of Forestry, 1 (2) (2000)
4. Yu.N. Neshataev, *Methods for the selection of plant associations* (L: Publishing house of St. Petersburg State University, 1971)
5. G.G. Sokolova, A.A. Kuznetsova, Geography and environmental management of Siberia, 16 (2013)
6. G.G. Sokolova, News of the Altai State University, 3–1 (83) (2014)

# The state of coenopopulations of the vegetatively-semimobile species of *Thymus* in Southern Siberia

*Evgeniya* Talovskaya[1*], *Irina* Barsukova[2]

[1]Central Siberian Botanical Garden of the SB RAS, 630090 Novosibirsk, Russia
[2]Katanov Khakass State University, 655017 Abakan, Russia

**Abstract.** The state of coenopopulations of vegetatively-semimobile dwarf shrubs *Thymus iljinii*, *T. minussinensis*, *T. mongolicus*, *T. petraeus*, widely distributed in steppe communities in Southern Siberia was studied. Adults of the species are a clump consisting of a primary and several partial bushes. For individuals characterized by the preservation of the main root, weak vegetative growth and reproduction, slow seizure of territory. The complex of data on organizational and population characteristics is analyzed. It is established that the real optimum state of coenopopulations of the studied species is achieved in the petrophytic variants of steppes located on the slopes of mountains. Conditions of high-mountain steppe and alpine plant communities are unfavorable for *Thymus*.

In the study of natural ecosystems and forecasting their development in recent years, researchers are increasingly using the concept of population organization of ecosystems. One of the approaches of the concept is the study of coenopopulations of species and their stability in a particular area [1, 2].

Species of *Thymus* L. are widely distributed in Southern Siberia. They are found in mountain and flat steppes, forest steppes and forests. As shown by our study, in Southern Siberia, the most *Thymus* species–are vegetatively-semimobile dwarf shrubs [3]. Adult individuals are characterized by the preservation of the main root, weak vegetative growth and reproduction, slow capture of the territory due to the close location of the partial structures to the primary bush. On the territory of the republics of Khakassia, Altai and Tuva we studied cenopopulations (CP) of four vegetatively-semimobile dwarf shrubs: *T. iljinii* Klok. et Schost. (CP 1), *T. minussinensis* Serg. (CP 2), *T. mongolicus* (Ronn.) Ronn. (CP 3-5), *T. petraeus* Serg. (CP 6-8). The majority of CP (1-3, 6-8) was studied in the conditions of petrophytic variants of steppes located in the middle or upper parts of the gravelly slopes or along the aligned peaks of the hills. The total projective cover (OPP) of grass in communities ranges from 55 to 70 %, the projective cover (PP) of *Thymus* from 4 to 7%. Coenopopulations 4 and 5 were studied in the alpine environment: CP 4 – in petrophytic version alpine meadows (OPP grass does not exceed 20 %, PP 1 %), the CP 5 in the condition petrophytic alpine steppe (OPP grass 55 %, PP 3 %). The state of the

* Corresponding author: kolegova_e@mail.ru

*Thymus* CP was assessed according to organism and population characteristics [4]. The diameter and phytomass of the bush of adult generative individuals, length of generative shoots, number of compound skeletal axes were chosen as organizational features; as population – ecological density of CP, phytomass of plants per m$^2$; the proportion of adult generative individuals in the CPU, the proportion of immature individuals and virginal states.

The assessment of the state of CP on the basis of organizational characteristics showed that their ecological optimum was in the petrophytic variant of steppe and was noted only in one CP 7 (Fig.). This CP is studied in the territory Nature reserve "Khakass", the area "Oglahty". The absence of anthropogenic action, leading to a violation of the integrity of individuals, increases the vitality of individuals and the values of almost all signs of the organism. However, competition of the species (ecological density of 38.9 ind./m$^2$, the CP 55, the projective cover of the rockslide with outcrops up to 25 %) in the condition of low available substrate, hinder the growth of individuals in the CP 7, the formation of new centres of consolidation, and maximize the values of their diameter. Individuals in the CP 1-3, 6, 8 are close to the optimal condition, they are located in the petrophytic steppes. Most of their parameters have average values. The conditions prevailing in the high-mountain steppe (CP 5) and in the alpine meadow (CP 4) are unfavorable for the development of individuals *Thymus*: all biometric parameters are sharply reduced.

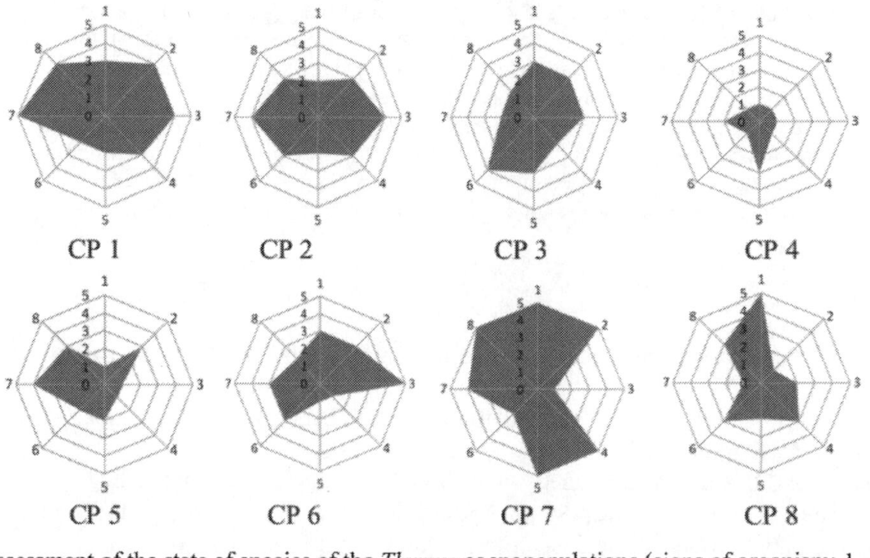


**Fig.** Assessment of the state of species of the *Thymus* coenopopulations (signs of organism: 1 – the number of skeletal axes, units; 2 – the length of generative shoots, cm; 3 – diameter of adult generative individuals, sm; 4 – phytomass in adult generative individuals, g; signs of population: 5 – phytomass species, g; 6 – proportion of adult generative individuals, %; 7 – the proportion of immature individuals and virginal, %; 8 – environmental CP density ind./m$^2$).

Analysis of population characteristics showed that in the most optimal conditions is CP 7, it is characterized by high rates of phytomass and ecological density, a large proportion of young individuals. The state of CP 1-5 is close to optimal due to a rather high proportion of im-v individuals (CP 1, 2, 5) and $g_2$ individuals (CP 3), as well as the phytomass of individuals (CP 3 and 4). In the CP 6, and 8 values of the majority of the population signs are dramatically reduced. This is due to different reasons. CP 6 is located at the top of the hill, where there is a constant blowing of snow in winter, lack of moisture in the summer. Under these conditions, there is a low viability of young individuals, reducing the duration of the adult generative state, which leads to a decrease in the ecological density and phytomass of individuals. In CP 8, the decrease in population characteristics is associated with the accumulation of dead grass (projective cover up to 20 %), which prevents the development of young individuals and their accumulation in the CP. Thus, the state of the CP 6 and 8 is approaching severe.

The coincidence of ecological and phytocenotic optima was noted only for one CP located in the petrophytic steppe in the reserve (CP 7). The condition of most of the CP is close to optimal (CPU 1-3). In the conditions of petrophytic variants of steppes on hilltops or with the accumulation of dead grass, along with the optimal state of individuals, the state of the CP is approaching to pessimal (CP 6 and 8). On the contrary, in the conditions developing in high-mountain plant communities, the state of individuals is pessimal at the optimal state of the CP (CP 4 and 5).

The analysis showed that the real optimum of vegetatively-semimobile shrubs of the *Thymus* in Southern Siberia is achieved mainly in the steppe zone of vegetation, in petrophytic steppes located on the slopes of the mountains. The conditions of high-mountain steppe and alpine plant communities in general are unfavorable for the studied *Thymus* species. The real optima of the organism and the population in such conditions do not coincide, and the values of the corresponding indicators are low.

The work was carried out with the financial support of the grant of RFBR within the framework of project № 18-04-00621 and State Assignment of Central Siberian Botanical Garden SB RAS № AAAA-A17-117012610053-9

## References

1. B. Walker, Conservation Biology, **9(4)** (1995).
2. L.B. Zaugol'nova, T.Yu. Braslavskaya, *Assessment and preservation of biodiversity of forest cover in European Russian nature reserves* (Moscow, Nauchnyy mir, 2000)
3. E.B. Talovskaya, V.A. Cheryomushkina, J. Ecosystem Ecology, **2(3)** (2017)
4. L.B. Zaugol'nova, Population structure of seed plants and problems of their monitoring: abstract of thesis of Doctor of Biological Sciences (St Peterburg University, Saint-Petersburg, 1994)

# Main trends of biodiversity dynamics after natural and anthropogenic "catastrophes" in spruce forests of the European part of Russia

*Nina* Ulanova*, and *Andrei* Kaplevsky

Lomonosov Moscow State University, Biological Dep., 119234 Moscow, Russia

**Abstract.** We analyzed the main trends of the change in the species richness of phytocenoses after catastrophic natural (beetle outbreaks, windfalls) and anthropogenic (clear cutting) disturbances. We examined the dynamics of the structural diversity of species richness of herb-dwarf scrubs layer with different reforestation technologies after the death of spruce stands. The study of similarity and ordination of vegetation showed the proximity of the undamaged forest to the unharvested stand, and the difference of these phytocenoses from clear-cut. The main determining factor of species richness was the intensity of phytocenosis disturbance after catastrophes. The reforestation dynamics of phytocenoses after catastrophic disturbances were analyzed. Clear cutting led to the formation of meadow communities with a sharp increase in the species and structural diversity of phytocenoses.

## 1. Introduction

The problem of conservation of forest biodiversity is one of the most important challenges in biology and ecology in 20-21 centuries and the central task of wildlife conservation. All anthropogenic (clear cutting, large area forest fires, industrial pollution) and natural (massive insect infestations, catastrophic windfalls) disturbances are considered to be negative factors leading to a reduction in biodiversity [1]. The large-scale death of forests leads to the reduction of diversity of biotopes, phytochors, and phytocenoses, the extinction of species and the reduction of their intraspecific genetic diversity. These conventional ideas are based on the comparison of data from different researchers, construction of spatiotemporal patterns, etc. However, identification of change patterns in species composition of communities can only be done by monitoring of the biodiversity of permanent sample plots in a series of phytocenoses along a relief gradient within a limited area (phytocatenae) for a long period of time after catastrophic disturbances [2].

We focus on an understudied aspect that is the dynamics of plant species composition (alpha-diversity) in the spruce forests of southern taiga and mixed coniferous-deciduous forests of the European part of Russia after "catastrophic" disturbances.

The natural environmental mechanism of the decay of spruce forest stand (which is the final stage of the dynamics of spruce phytocenoses at the final stage of succession in the

* Corresponding author: NUlanova@mail.ru

European part of Russia) is implemented in catastrophic windfalls, fires or dead-wood patches during outbreaks of eight-dentated bark beetle (*Ips typographus* L.).

## 2. The intensity of disturbances in phytocenoses after natural and anthropogenic disasters

Catastrophic natural phenomena that cause the death of spruce forests create disturbances that vary in scale (Fig. 1). During fires, a significant part of tree stands and under-growing vegetation dies, and there is a very wide range of soil damage. In the case of catastrophic windfalls, the scale of the loss of the stand and the ground cover varies, and the soil cover damage is minor [3]. With partial preservation of the forest stand and undergrowth in windfalls, only a redistribution of dominant species with a slight change in the species composition occurs in the herb-dwarf shrubs layer [2, 4]. In areas of spruce drying during outbreaks of the eight-dentated bark beetle, the soil and ground cover remains mostly untouched, but the proportion of dead spruce varies from 0 to 100 %. The degree of ecosystem disturbance during disasters that lead to the destruction of spruce forests determines the rate of vegetation restoration in areas of fire-damaged forest, windfalls, and in the epicenters of spruce drying.

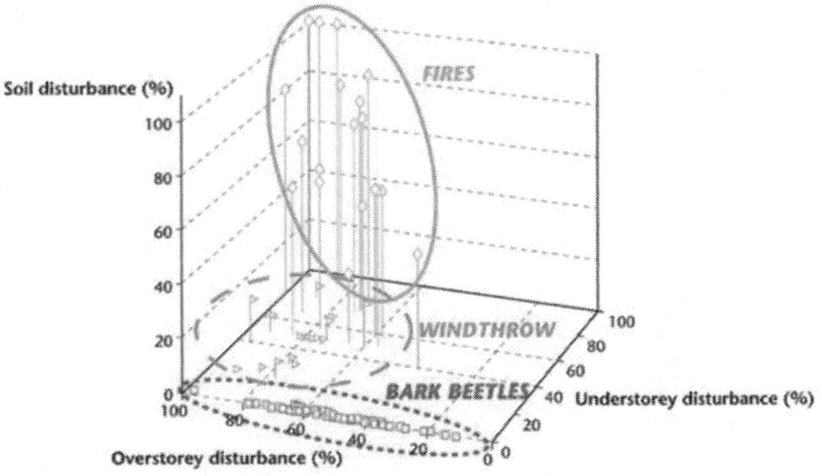


**Fig. 1.** The intensity of the destruction of the forest stand, ground cover and soil in phytocenoses after natural catastrophes (fire, massive windfall, outbreaks of bark beetle) [5].

## 3. Increase of biodiversity: species richness and structural diversity

Natural and anthropogenic disasters lead to transformations of the original phytocenoses of different intensity. As a result, an increase in biodiversity in new communities occurs; in areas of dead-wood spruce it is insignificant, and it is higher in areas of catastrophic windfalls. With the complete destruction of spruce forest stand (not only the dead trees) in the course of clear cutting, a radical change in the soil cover [6] and transformation of forest communities into grass and shrub communities occurs, which leads to a fundamental change in the vegetation cover. In newly formed meadow communities, biodiversity increases dramatically due to emergence of non-forest species [2]. Additionally, a meta-analysis of the results of 238 studies of biodiversity in various forests of the world revealed an increase in the species richness of vascular plants after catastrophic windfalls and death of the forest stand as a result of outbreaks of bark beetles and logging [7].

Let us consider changes in vegetation in areas of spruce drying in 2012 in the western part of the Moscow Region (Zvenigorod Biological Station of Moscow State University) as an example.

In 2013, we established three permanent sample plots of the same size (800 $m^2$) in a spruce-weaselsnout forest: one in an area with the spruce forest stand that died in 2012 (bark beetle infection), one in an area of clear cutting of dead-wood in the winter of 2012–13, and one in an area with living spruce tree stand (control). Studies were conducted in August 2014, 2015 and 2016 using a uniform methodology. Three transects 40 m long and 40 cm wide each were made on the sample plots. The frequency of species in the herb-dwarf-shrubs layer was studied at each meter of the transect, using a rigid frame 40 x 100 cm in size, divided into 20 x 20 cm squares by removable threads.

In spruce dead-wood forest, species retained their dominance in the herb-dwarf shrubs layer (Fig. 2). After snagging, the floristic composition of the herb-dwarf-shrub layer increased by a factor of 2, compared to the undisturbed spruce stand, as a result of the death of the herb-dwarf-shrub layer, disturbances of moss and soil cover during the wood transportation, burning of cuttings and subsequent introduction of new species. Other species became dominant. As a result of significant soil disturbances, patchiness of the herb-dwarf-shrub layer became very high.

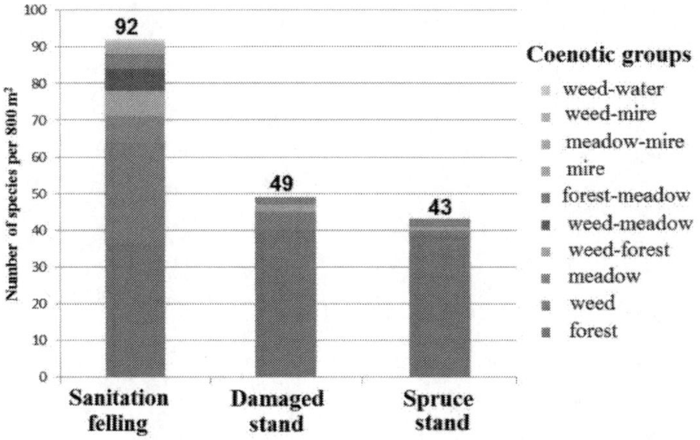


**Fig. 2**. The coenotic structure of the species richness of the herb-dwarf-scrub layer in the spruce-weaselsnout stand (control), in the second year after the death of the forest stand with the sanitation felling of dead-wood and with the preservation of the dead-wood (damaged stand).

The coenotic spectrum of the herb-dwarf-shrub layer in the spruce forest after the death of spruce trees corresponds to the spectrum of the original forest (Fig. 2). During the second year, there was a change in the frequency of species in the herb-dwarf-shrub layer and species of new coenotic groups characteristic of the original forest were introduced. The frequency of moss cover species decreased due to shading from growing hazel shrubs. In the slash area, the increase in the number of coenotic groups was doubled due to the introduction of herbaceous species and mosses that were not characteristic for the original community. In the slash area, the proportion of forest species significantly decreased, and the proportion of weed, meadow and weed-meadow species increased. The type of the phytocenosis of the slash areas can be characterized as forest-meadow.

## 4. Regenerative dynamics of phytocenoses after catastrophic disturbances

Significant disturbances of phytocenoses and soils lead to the following types of successions: demutations, secondary incomplete and quasi-primary. Changes in vegetation depend on the initial forest type. Thus, in the areas of drying of spruce trees in the Moscow Region, in spruce-blueberry and spruce-blueberry-shamrock forests, fluctuations go through a stage with rowanberry forest with spruce undergrowth. In the spruce-weaselsnout and complex spruce forests, demutation goes through a stage of lime and maple forests with spruce undergrowth. In complex spruce forests with hazel, the incomplete secondary succession ends with a hazel shrub forest.

An alternative method of forest management (preservation of the dead stand and natural regeneration of the forest) is possible only in forests that have a protected status. The preservation of dry stands and wind-fall areas in spruce forests leads to the natural course of reforestation, which preserves the forest phytocenoses and changes only the proportions of dominant species in the forest stand. As a result, mixed stands with broad-leaved species form, which are highly resistant to pests and forest diseases. Forests with a complex structure replace monocultures of spruce, which contributes to the restoration of the diversity of forests characteristic of the coniferous-deciduous forest zone. Such natural forests are most likely characteristic of the coniferous-deciduous forest zone.

Forest management in spruce forests requires clear sanitation felling of the dead tree stand in the event of bark beetle outbreaks, as well as the clearing of catastrophic windfall and fire sites. The massive use of clear cuttings over the last ten years has led to an increase in the slash area where meadow communities have formed. In new meadow communities the biodiversity increases dramatically due to the presence of non-forest species [2]. From the point of view of a biologist, this process cannot be considered negative for nature. If our ideology requires an increase in biodiversity, then the formation of meadow vegetation is the best solution. Several publications that came out in recent years raise the question of the importance of the first succession stages for increase of forest biodiversity [7, 8, 9, 10].

## References

1. A.S. Isaev, *Monitoring of biological diversity of Russian forests: methodology and methods* (M, Science, 2008).
2. N.G. Ulanova, Herald of TVGU. Series: Biology and Ecology, **3** (2018)
3. N.G. Ulanova, For. Ecol. Manag., **135** (2000).
4. N.G. Ulanova, O.V. Cherednichenko, Izvestia of Samara Sc.C. RAS, **14** (2012)
5. P.J. Burton, BC J. Ecos. and Manag., **9** (2008)
6. A.A. Dymov, Eurasian Soil Science, **7** (2017)
7. S. Thorn, C. Bässler, R. Brandl, P.J. Burton, R. Cahall, J.L. Campbell, J. Castro, C.Y. Choi, T. Cobb, D.C. Donato, E. Durska, J.B. Fontaine, S. Gauthier, C. Hebert, T. Hothorn, R.L. Hutto, E.J. Lee, A.B. Leverkus, D.B. Lindenmayer, M.K. Obrist et al., J. Appl. Ecol., **55** (2018).
8. M.E. Swanson, J.F. Franklin, Ch.M. Crisafulli, D.A. Della Sala, R.L. Hutto, D.B. Lindenmayer, Fr.J. Swanson, Front. Ecol. Environ., **9** (2011).
9. D.P. Blair, L.M. McBurney, W. Blanchard, S.C. Bank, D.B. Lindenmaye, Ecological Appl., **26** (2016).
10. P.J. Fornwalt, Ch.C. Rhoades, R.M. Hubbard, R.L. Harris, A.M. Faist, W.D. Bowman, For. Ecol. Manag., **409** (2018)

# Various methods for isolating DNA from fruiting bodies on the example of Pleurotus pulmonarius (higher basidiomycetes) of the Novosibirsk Region for barcoding of edible and medicinal mushrooms

*Vyacheslav* Vlasenko*, *Sergey* Asbaganov and *Anastasia* Vlasenko

Central Siberian Botanical Garden SB RAS, 630090 Zolotodolinskaya st., 101, Novosibirsk, Russia

**Abstract.** We conducted experiments to increase the concentration of extracted DNA from the fruiting bodies of this species using various lysis buffers and extraction time. In general, DNA isolation according to the protocol using SDS made it possible to obtain higher concentrations thereof, as in the case of increasing the extraction time from 1 to 24 hours and using additional purification with chloroform.

## 1 Introduction

*P. pulmonarius* (Fig. 1) have medicinal properties, the polysaccharides contained in them exhibit a high antioxidant, antitumor, immunomodulatory, antiviral effects [1, 2].

DNA barcoding is a potent approach for rapid identification of fungal specimens, it can be used to accurately identify edible and medicinal mushrooms.

In this study, experiments were conducted on the isolation of DNA from the fruit bodies of oyster mushrooms to obtain its highest concentration.

## 2. Materials and methods

Fruiting bodies of fungi were collected in the field studies in various habitats of the Novosibirsk Region. Dried specimens were preserved in the herbarium, they were used for molecular genetic studies.

An overview of taxa used for studies, shows the species names, herbarium voucher, habitats and substrates given in Table 1.

DNA extraction and sequencing

Specimens of *P. pulmonarius* was used for molecular analysis. A fragment of fungal fruiting bodies (20 mg) was homogenized in lysis buffer and extract the DNA with NucleoSpin® Plant II kit was used. The ITS1-5.8S-ITS2 region of the rDNA was amplified by PCR with the primers ITS1F and ITS4B (Table 2).

* Corresponding author: vlasenkomyces@mail.ru

**Fig. 1.** *Pleurotus pulmonarius* – fruiting bodies in nature on *Sorbus sibirica* and mycelium in culture on agar medium in Petri dish.

**Table 1.** Pleurotoid fungi used for studies

| Species | Herbarium voucher | Habitat | Substrate |
|---|---|---|---|
| *Pleurotus pulmonarius* | NSK 1014214 | birch forest | *Sorbus sibirica* |
| *Pleurotus pulmonarius* | NSK 1014215 | birch forest | *Betula pendula* |

**Table 2.** Primers used for studies

| Loci | Primers | (F/R) | Sequence (5′-3′) |
|---|---|---|---|
| ITS1-5.8S-ITS2 | ITS1-F | F | CTTGGTCATTTAGAGGAAGTAA |
| ITS1-5.8S-ITS2 | ITS4-B | R | CAGGAGACTTGTACACGGTCCAG |

For PCR, HS Taq DNA Polymerase (Evrogen, Russia) was used. PCR reactions were performed in a C1000 Thermal Cycler (Bio-Rad, USA). PCR results were checked at Gel Doc XR+ Imager (Bio-Rad, USA). DNA amplicons sequencing performed in SB RAS Genomics Core Facilities (Novosibirsk, Russia).

Sequences were aligned using ClustalW methods. The ITS sequences were aligned in MEGA 7. Phylogeny reconstruction was inferred using the UPGMA method. The percentage of replicate trees in which the associated taxa clustered together in the bootstrap test (100 replicates) is shown to be next to the branches. The evolutionary distances were computed using the Maximum Composite Likelihood method and were in the units of the number of base substitutions per site. The differences in the composition bias among sequences were considered in evolutionary comparisons. All ambiguous positions were removed for each sequence pair (pairwise deletion option). There were a total of 609 positions in the final dataset. Evolutionary analyses were conducted in MEGA 7.

## 3. Results and Discussion

Lysis buffers PL-1 (CTAB method) and PL-2 (SDS- method) were used. Two types of extraction were used: 1 hour according to the manufacturer's protocol and 24 hours with additional purification with chloroform. The results of measurements of the concentration of extracted DNA are presented in Table 3.

**Table 3.** The concentration of DNA isolated from the studied samples

| Samples | 1 | 2 | 1 | 2 |
|---|---|---|---|---|
| Method | CTAB | CTAB | SDS | SDS |
| C (extraction 1 h) | 24 | 27 | 29 | 39 |

| | | | | |
|---|---|---|---|---|
| C (extraction 24 h) | 172 | 94 | 38 | 106 |

**Note**: C (concentration) ng/µl. Samples: 1 – *Pleurotus pulmonarius* NSK 1014214, 2 – *P. pulmonarius* NSK 1014215. Optical density ratios A260/A280 = 1.84-1.97.

The results of PCR imaging are presented in Fig. 2.

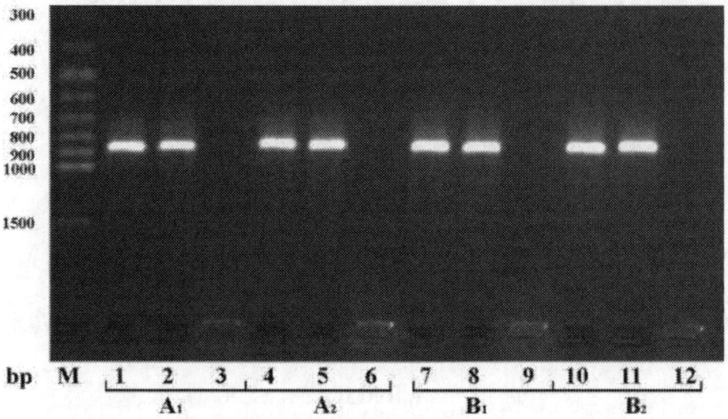


**Fig. 2.** Visualization of PCR results. CTAB – method: A (samples 1-6). A1 – manufacturer protocol extraction, A2 – extraction with an increase in extraction time. SDS – method: B (samples 7-12). B1 – manufacturer protocol extraction B2 – extraction with an increase in extraction time. Samples DNA: 1, 4, 7, 10 – Pleurotus pulmonarius NSK 1014214. Samples DNA: 2, 4, 8, 11 – Pleurotus pulmonarius NSK 1014215. Samples DNA: 3, 6, 9, 12 – *Meriderma carestiae* myxomycete DNA was used for fungal-specific primer monitoring).

Phylogenetic analyses

Additional 2 ITS sequences of other Pleurotus species based on BLAST results and 4 ITS sequences of other species were retrieved from GenBank (http://www.ncbi.nlm.nih.gov/Genbank/). We first generated a 2 new sequence for the ITS1-5.8S-ITS2 region for *P. pulmonarius*. The final dataset consisted of 7 ITS sequences. An overview of all taxa and on sequences used for tree reconstruction, shows the species names, herbarium vouchers/strain and Genbank accession numbers given in Table 4.

**Table 4.** Sequences used in alignment

| Species | Herbarium voucher/Strain | Genbank accession numbers |
|---|---|---|
| *Pleurotus pulmonarius* | NSK 1014214 | MN179418 |
| *Pleurotus pulmonarius* | NSK 1014215 | MN179421 |
| *Pleurotus pulmonarius* | FPPMK-L | JX429930 |
| *Pleurotus 'sajor-caju'* | H-1 | JQ837470 |
| *Pleurotus* cf. *eryngii* | C1 | FJ514549 |
| *Pleurotus ostreatus* | 6689 | AY450345 |
| *Pleurotus populinus* | 9936 | AY450346 |

The molecular phylogenetic analyses placed the specimens of *Pleurotus* genus from Novosibirsk Region close to *P. pulmonarius* (Fig. 3).

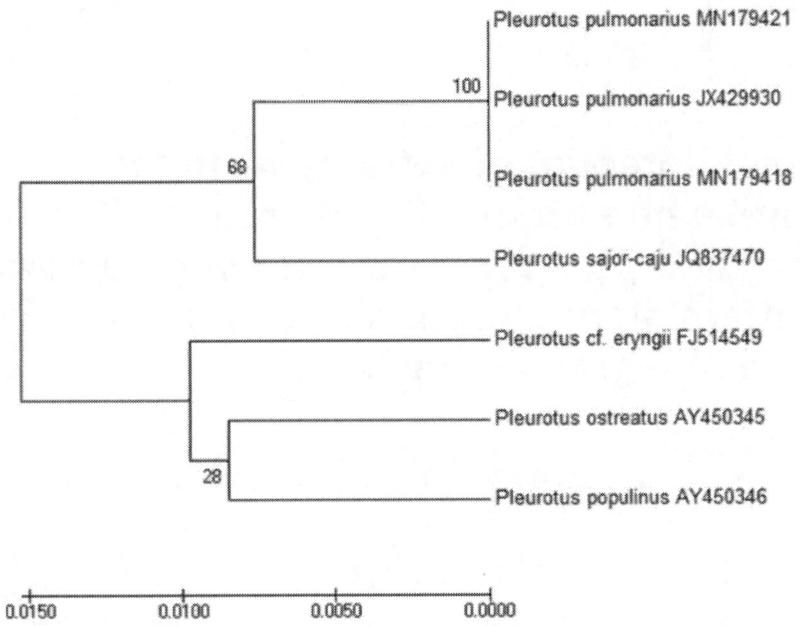


**Fig. 3.** UPGMA trees based on ITS sequences showing the phylogenetic relationships between the *P. pulmonarius* species from Novosibirsk Region and other related species of *Pleurotus* genus. Herbarium specimens from the Bioresources scientific collection (USU 440537) MG Popov Herbarium (NSK) were used.

## 4. Conclusions

DNA isolation according to the protocol using SDS allows to obtain higher concentrations, as and in the case of increasing the extraction time from 1 to 24 hours and using additional purification with chloroform.

## Acknowledgements

The reported research was funded by Russian Foundation for Basic Research and the government of the Novosibirsk Region of the Russian Federation, grant № 18-44-543018 R_Mol_a.

## References

1. T.V. Teplyakova, N.V. Psurtseva, T.A. Kosogova, N.A. Mazurkova, V.A. Khanin, V.A. Vlasenko. Int. J. Med. Mushr., **14(1)** (2012)
2. V.A. Vlasenko, A.V. Vlasenko. BIO Web Conf., **IV**, 00044, 11 (2018)

# Medicinal potential of extracts from the chanterelle mushroom, Cantharellus cibarius (Review) and prospects for studying its strains from differs plant communities of ultra-continental regions of the Asia

*Vyacheslav* Vlasenko[1*], *Dejidmaa* Turmunkh[2], *Enkhtuya* Ochirbat[3], *Dondov* Budsuren[2], *Kherlenchimeg* Nyamsuren[3], *Javkhlan* Samiya[3], *Burenbaatar* Ganbaatar[3], and *Anastasia* Vlasenko[1]

[1]Central Siberian Botanical Garden SB RAS, 630090 Zolotodolinskaya st., 101, Novosibirsk, Russia.
[2]Plant Protection Research Institute MULS, 17024, Khoroo 11, Khan-Uul district, Ulaanbaatar,
[3]Institute of General and Experimental Biology MAS, Peace Avenue-54b, Bayanzurkh District, Ulaanbaatar-13330, Mongolia

**Abstract.** The article reviews medicinal properties of *Cantharellus cibarius*. Fruiting bodies of this species are a promising source of extracts with vitamins, carbohydrates, proteins, polysaccharides, sterols, phenolic acids, which various biological properties.

## 1 Introduction

Mushrooms are becoming more and more important in our diet for their nutritional characteristic. Basidiomycetes contain a wide range of various drugs such as polysaccharides, organic acids, lipids, steroids, tetracyclic triterpenes, which are of interest for medical applications [1, 2].

One of the most frequently harvested mushrooms in forests of ultra-continental regions of the Asia is Yellow chanterelle, *C. cibarius* (fig. 1) is an edible mushroom with medicinal value.

Chanterelle is a beneficial nutrient for human health with high levels of vitamins B and C, contains a great number of carbohydrates and proteins and a low amount of fat, phenolic compounds and organic acids [3].

This species is appreciated because of the aroma and taste of its fruiting bodies. *C. cibarius* is one of the most important wild edible mushrooms, therefore, the assessment of its biological properties is of great interest to know its potential as an alternative treatment to chemopreventive strategies when it is consumed as part of a diet.

## 2 Materials and methods

* Corresponding author: vlasenkomyces@mail.ru

An up-to-date review based on published data presents the research of main medicinal properties of extracts from fruiting bodies and mycelium of *C. cibarius*. Information on biological activity of *C. cibarius* is given in chronological order.

**Fig. 1.** Fruiting bodies of *Cantharellus cibarius* in nature.

## 3 Results and Discussion

*Cantharellus cibarius* has been reported to display a wide variety of biological properties, including antimicrobial, cytotoxicity, antioxidant, antihypoxic, antihyperglycemic, wound healing, anti-inflammatory, iron-chelation activity.

Ethyl acetate, acetone, chloroform and ethanol extracts of *C. cibarius* were tested for antimicrobial activity. Antimicrobial activity against some Gram (+) and Gram (-) bacteria, yeasts, filamentous fungi and actinomycetes was revealed [4].

Many natural foods (including antioxidants) with anti-cancer and anti-inflammatory activity inhibit NF-*k*B. The inhibitory activities of the extracts of *C. cibarius* and isolated compounds were investigated in an enzyme-based ELISA NF-*k*B assay. Of the tested compounds, ergosterol, ergosterol peroxide and cerevisterol were noted to have the most potent inhibition of NF-$\kappa$B activation [5].

Antioxidant activity and bioactive compounds from *C. cibarius* was studied. High phenol and flavonoid content with promising antioxidant properties was shown. Methanolic extracts of wild mushroom *C. cibarius* showed significant antimicrobial activity against *B. subtilis*, *S. aureus*, *E. coli*, *P. aeruginosa* and *Candida albicans* [6].

Anti-inflammatory properties related to the activity of phenolic compounds originating from edible mushroom *C. cibarius* were tested in vitro by using LPS-activated RAW 264.7 macrophages. Experiments demonstrated the influence of extracts on the expression of inflammation markers, such as IL-1$\beta$ and IL-6, and the production of nitric oxide (NO) [7].

Fatty acids contained in mushrooms are capable of supporting anti-inflammatory processes in humans owing to their high content of unsaturated fatty acids. Extracts of *C. cibarius* demonstrated the presence of fatty acids exhibiting antagonist activity toward receptors activated by peroxisome proliferator-activated receptor $\gamma$ (PPAR$\gamma$). PPAR$\gamma$ agonists inhibit the development of insulin resistance, thus being an efficient anti-diabetic element, and also demonstrate a therapeutic effect in the case of diseases associated with inflammation and certain tumors [8].

Phenolic acids possess valuable biologically properties: anti-inflammatory, antioxidant, anticarcinogenic and others. One of the most active antioxidants found in mushrooms, caffeic acid, was found in *C. cibarius*. It exhibits both anti-inflammatory and antioxidative activities [9, 10].

Water extracts of *C. cibarius* were studied. Comet assay was used to evaluate the antigenotoxic properties of several concentrations of aqueous extracts. As a test system human mononuclear cells exposed to methyl methanesulphonate (MMS) in vitro were used. Extracts showed a decrease in genotoxic properties of MMS [11].

*C. cibarius* extract is valuable source of antihypoxic activity. Extract showed protective effect against hypoxia in some models. In particular, it produced a significant and dose-dependent antihypoxic effect in the hemic and circulatory hypoxia models. Pharmacological effects attributable to the presence of polyphenols in the extract [12].

Methanolic and ethyl acetate extracts of *C. cibarius* exhibit potency to chelate excessive iron in mice. Iron-overloaded animals treated with the extract, showed a dramatic decrease in plasma iron content when compared with the control group. The highest activity was observed in the methanolic extract [13].

The inhibitory effect of pure liquid cultures of *C. cibarius* was tested in vitro against 16 pathogenic Gram-positive and Gram- negative bacterial strains, and 4 strains of oval fungi of the genus *Candida*. Significant inhibitory effect of the culture fluids of *C. cibarius* against all tested microbial species was determined. MIC50 of the cultures were similar and lower than those of control broad-spectrum antibiotics thiamphenicol and gentamicin. Liquid cultures of *C. cibarius* can be used to produce effective antimicrobial drugs with broad spectrum activity against bacteria and fungi [14].

Various fractions of the methanolic extract including n-hexane, chloroform, ethyl acetate, n-butanol, and water, were evaluated for antioxidant activity. The n-hexane fraction had the highest amounts of flavonoids contents, nitric oxide scavenging activity, and total phenols. However, it was the ethyl acetate fraction that showed the highest DPPH scavenging activity, although the aqueous fraction displayed higher reducing power [15].

The antioxidant, antihyperglycemic and antimicrobial effects of both ethanolic and hydromethanolic extracts of the fruiting bodies of *C. cibarius* were evaluated. Ethanolic extract showed high ferrous ion chelating, 15-lipoxygenase and αglucosidase inhibitory activities. Ethanolic extracts were more active against *Staphylococcus aureus* ATCC 25923 than the hydromethanolic ones [16].

*C. cibarius* was analyzed for their fatty acid compositions and antioxidant capacities. Very high metal chelating activity was found. Mushrooms have a potential to be a natural antioxidant in food industries as antioxidant agent [17].

The experimental data obtained in rats show that the extract of *C. cibarius* showed a significant wound healing and anti-inflammatory effect, which can become a scientific justification for the medicinal use of the golden chanterelle mushroom in the treatment of wounds [18].

Activities of cyclohexane, dichloromethane, methanol and aqueous extracts of *C. cibarius* were tested. Broth microdilution assay was performed against 10 bacterial, with emphasis on *Helicobacter pylori*. Methanol extract was the most active against *H. pylori* strains with minimal inhibitory concentration values between 4 and 32 $\mu$g/mL. All extracts were active against antibiotic resistant *H. pylori*. Chemical analysis has shown the presence of linoleic, *cis*-vaccenic, and oleic acids, sterols, $\beta$-glucans, and polyphenolic compounds [19].

The extracts of *C. cibarius* were used for the evaluation of their effects on cell viability of four cancer cell lines (A172 glioblastoma, U87 glioblastoma, CH157-MN meningioma and PAN-C1 pancreatic) and one NIH3T3 fibroblast cell line. Treatments with high concentrations of *C. cibarius* extracts resulted in significantly reduced number of viable cancer cells for all cell lines tested [20].

## 4 Conclusions

Medicinal properties of *C. cibarius* is of great interest to know its potential as an alternative treatment to chemopreventive strategies when it is consumed as part of a diet. The data on the antiviral activity of chanterelles are absent. Therefore, we plan to study the antiviral activity of extracts from fruiting bodies of *C. cibarius* from ultra-continental regions of the Asia (Russia and Mongolia), in which this species is a valuable food resource.

The reported research was funded by Russian Foundation for Basic Research, grant № 19-54-44002 Mong_T.

## References

1. T.V. Teplyakova, N.V. Psurtseva, T.A. Kosogova, N.A. Mazurkova, V.A. Khanin, V.A. Vlasenko, Int. J. Med. Mushr., **14(1)** (2012)
2. V.A. Vlasenko, A.V. Vlasenko, BIO Web Conf., **IV** (2018)
3. B. Muszyńska, K. Sułkowska-Ziaja, H. Ekiert, Acta Scient. Polon. Hortor. Cult., **12** (2013)
4. B. Dulger, A. Gonuz, F. Gucin, Pakistan J. Biol. Sciences, **7(9)** (2004)
5. J.A. Kim, D. Tay, E.C. Blanco, Phytotherapy research, **22(8)** (2008)
6. C. Ramesh, M.G. Pattar, **2(2)** (2010)
7. Palacios, M. Lozano, C. Moro, M.D'Arrigo, M.A. Rostagno, J.A. Martinez, A. Villares, Food Chemistry, **128** (2011)
8. S.S. Hong, J.H. Lee, W. Jeong, N. Kim, H.Z. Jin, B.Y. Hwang, D. Lee, Bioorg. Med. Chem. Lett, **22** (2012)
9. C. Mendez-Espinoza, E. Garcia-Nieto, A.M. Esquivel, M.M. Gonzalez, E.V. Bautista, C.C. Ezquerro, L.J. Santacruz, Int. J. Med. Mushrooms, **15(3)** (2013)
10. F.S. Reis, A. Martins, L. Barros, I.C.F.R. Ferreira, Food Chem. Toxicol., **50** (2012)
11. C. Mendez-Espinoza, E. Garcia-Nieto, A.M. Esquivel, M.M. Gonzalez, E.V. Bautista, C.C. Ezquerro, L.J. Santacruz, Int. J. Med. Mushrooms, 15(3) (2013)
12. M. Khalili, M.A. Ebrahimzadeh, F. Omrani, M. Karami, **16(4)** (2014)
13. M. Khalili, M.A. Ebrahimzadeh, M. Kosaryan, A. Abbasid, M. Azadbakhta, RSC Advances, **5** (2015)
14. T.P. Popova, Int. J. Curr. Microbiol. App. Sci, **4(3)** (2015)
15. M.A. Ebrahimzadeh, Y. Safdari, M. Khalili, Int. J. Med. Mushrooms, **17** (2015)
16. D.E. Zavastin, A. Bujor, C. Tuchiluş, C.G. Mircea, S.P. Gherman, A.C. Aprotosoaie, A. Miron, J. Plant Develop., **23** (2016)
17. Türkekul, F. Çetin, M. Elmastaş, J. Appl. Biol. Chem., **60(1)** (2017)
18. D. Nasiry, A.R. Khalatbary, M.A. Ebrahimzadeh, Int. J. Med. Mushrooms, **19(10)** (2017)
19. M. Kolundžić, T. Stanojković, J. Radović, A. Tačić, M. Dodevska, M. Milenković, F. Sisto, C. Masia, G. Farronato, V. Nikolić, T. Kundaković, J. Med. Food., **20(8)** (2017)
20. K.L. Chin, S. Khanal, G.R. Hankins, V. Perla, Y. Qi., Act. Scient. Agriculture, **2(5)** (2018)

# Improvement of fire danger rating and vegetation fire behaviour prediction on protected areas

*Aleksandra* Volokitina[1,*], *Dina* Nazimova[1], *Tatiana* Sofronova[2], and *Mikhail* Korets[1]

[1]Sukachev Institute of Forest SB RAS, ul. Akademgorodok, 50/28, 660036 Krasnoyarsk, Russia
[2]Astafiev Krasnoyarsk State Pedagogical University, str. A. Lebedevoy, 89, 660049 Krasnoyarsk, Russia

**Abstract.** Protected areas (PAs) are established to conserve biological diversity, to maintain nature complexes and objects in their natural condition. Strict nature reserves prevail in Russia by their total area. The whole nature complex is forever extracted from economic use in nature reserves. Here it is prohibited to pursue any activity which might disturb or damage the nature complexes. Even under the existing strict protection from anthropogenic ignition sources, vegetation fires do occur on their territory. Besides, lightnings – these natural ignition sources – are impossible to exclude. Since large destructive fires are impermissible in nature reserves, the later especially need vegetation fire behavior prediction for fire management. Fire behavior prediction includes fire spread rate, development (from surface fire into crown or ground one) and fire effects. All this is necessary for taking optimal decisions on how to control each occurring fire and how to suppress it. The Sukachev Institute of Forest SB RAS has developed a method to improve forest fire danger rating and a technique of vegetation fire behavior prediction using vegetation fuel maps (VF maps).

## 1 Introduction

Protected areas (PAs) are established to conserve biological diversity, to maintain nature complexes and objects in their natural condition. Taking into account the special regime and status of the environmental conservation institutions there, various IUCN categories of PAs are distinguished: strict nature reserves or wilderness areas (including biosphere ones), national parks, habitat/ species management areas, natural monuments, etc. Strict nature reserves prevail in Russia by their total area.

Large destructive fires are impermissible to develop in nature reserves and in natural monuments. This issue can be resolved largely by improvement of fire danger rating, fire behavior prediction and fire management. Vegetation fire management in nature reserves is quite possible on the basis of modern fundamental fire science developments which improve fire danger rating, prediction of an occurred fire depending upon weather

* Corresponding author: volokit@ksc.krasn.ru

conditions, and make it possible to control flame (or flameless) combustion spread over the area and to take optimal decisions on fire suppression.

## 2 Background

Analysis of vegetation fires in Russian strict nature reserves was carried out by L.V. Kuleshov and V.N. Korotkov [1]. The dynamics of fires was analyzed by them for the period from 1975 to 2000. They revealed a general trend of growth in the number of fires.

There are no special recommendations on fire protection at PAs in Russia. Until now, during their forest inventory, developments intended for forest management activities are used, for example, forest fire maps are created, in which categories of vegetation plots are divided into classes of natural fire danger [2]. This led to the fact that not only dry lichen pine forests, but in general all the debris-strewn areas (fire-damaged forests, dead-standing forests) and all coniferous young growth with any forest floor were included in the first class of the scale. And not only poorly burned sphagnum and long-moss pine forests, but also all grass forest types are classified as the fourth fire hazard class, although in spring and autumn in southern Siberia, fires in grass types with a predominance of grasses and sedges often turn into a natural disaster. In addition, there was no quantitative pyrological characteristics of areas included in one class.

Studies of quantitative criteria for the rate of drying to an ignitbale state in various categories of forest and non-forest areas in different regions were actively carried out in Russia in the 1960-1970s according to the methods of Professor N.P. Kurbatsky [3]. Scales of priority in ignition of plots were drawn up with an indication of the value of the forest fire drought index, at which plots of this category reach an ignitable state. But the total number of possible categories of plots in Russia is too large to cover all of them with research. Moreover, the forest floor of many forest types, identical in name, had large pyrological differences. It is almost impossible to study individually all this great variety of categories of plots. Their pyrological classification was necessary, the division of all forest floors into a number of pyrological types.

The analysis of published observations on drying of forest floors to an ignitable state, as well as special studies on the layer-by-layer drying of forest floors [4], allowed us to develop further the classification of forest fuels by N.P. Kurbatsky [5, 6]: in each fuel group we distinguished fuel types [7]. The division of the first fuel group (including mosses, lichens, and small plant residues) into types made it possible to show them on vegetation fuel maps (VF maps), the methods for creating which have now been developed. VF types of this group were named as the primary fire carriers (PFC) and their pyrological characteristics were obtained on the basis of many years of research on the moistening, drying and burning of various forest floors in different regions [7].

## 3 Improvement of fire danger rating on PAs

Among the fire hazard assessments used in forest fire practice for PAs, the most relevant is fire danger rating according to weather conditions and natural fire hazard assessment. The Sukachev Institute of Forest SB RAS developed methodological recommendations aimed at improving these types of fire danger rating. Thus, in order to improve the fire weather danger rating, it is proposed to use an improved index that takes into account the humidity and hygroscopicity of vegetation fuels and, unlike previous indices, “works” at negative temperatures. It is recommended to use this index in northern and central taiga areas with a predominance of moss and lichen in the forest floor, as well as in the Zabaikalie and Far

East regions, where cured grass fires often occur in spring and autumn at below-zero temperatures [8, 9].

To improve the assessment of the natural fire hazard, it was proposed to use VF maps, the creation methods of which are based on the detailed VF classification [7]. To date, information databases have been compiled in GIS and VF maps have been made for a number of strict nature reserves: Stolby, Sayano-Shushensky, Kuznetsky Alatau, Ubsunurskaya hollow.

## 4 Vegetation fire behaviour prediction on PAs

VF maps also make it possible to predict the behavior of the fires that have occurred, since the maps contain the necessary pyrological characteristics of the vegetation. Under the behavior of a vegetation fire we imply the speed of its spread over the territory, the type of fire, its development (transition from a surface to crown or ground fire), as well as the immediate fire effects in the form of tree mortality, depending on the intensity of burning, the tree species and its average diameter [10]. Software programs for VF mapping and vegetation fire behavior prediction have been developed and patented, and they have been retrospectively verified and showed good performance [11, 12].

## 5 Vegetation fire management on PAs

To manage vegetation fires on PAs, first of all, it is necessary to create an information database containing VF maps. Since Russian PAs go through the first forest inventory type, it is quite possible to create the information database using the data of the previous forest inventory based on the technique developed by the Sukachev Institute of Forest SB RAS for analysis of forest type schemes and their pyrological characteristics. If old forest inventory data are not relevant already and a new forest inventory is being planned then it would be better to create a new information database for making VF maps with more accurate pyrological characteristics on the basis of the Sukachev Institute of Forest identifier for primary fire carriers in the process of forest inventory.

Further, on the basis of a chosen fire spread model, the developed computer software is used to predict behavior of a specific fire: specific pyrological characteristics are taken from the VF map as well as weather data on a given time. The software program calculates the surface fire edge spread in given time periods. On the sites where a surface fire can develop into a crown or a ground one, the program shows this probability. Besides, the software reflects possible fire effects in forest stands depending upon the combustion intensity, tree species and mean tree stand diameter in terms of tree mortality expressed in percent. Additionally, taking into account the length of a fire edge, the software calculates the necessary suppression forces for a given stage. By now, the vegetation fire behavior prediction software has successfully gone through a retrospective performance test in plain conditions (the Chunsky Forest Office of the Krasnoyarsky krai) and in mountain conditions (the Stolby Nature Reserve of the Krasnoyarsk krai).

## 6 Conclusion

Vegetation fire management is quite realistic on Pas of Russia. Most PAs have the necessary information database because their forest inventory is of a high forest inventory type. Nature reserves have their own weather stations and use automatic ones. The software for vegetation fire behavior prediction is pretty straightforward for the specialists of nature reserves to master.

This paper is prepared under the support of the Russian Foundation of Fundamental Research grant № 18-05-00781A.

## References

1. L.V. Kuleshova, V.N. Korotkov, *Monitoring of communities on burnt areas and fire management in strict nature reserves* (VNIIprirody, Moscow, 2002)
2. *Collection of agency-level acts on forestry* (Gosleskhoz, Moscow, 1979)
3. N.P. Kurbatsky, *Methodical guidelines for practical development of local fire danger scales* (TsNIILKh, Leningrad, 1954)
4. A.V. Volokitina, Lesovedenie, **1** (1985)
5. N.P. Kurbatsky, *Methods and strategies of forest fire fighting* (Goslesbumizdat, Moscow, 1962)
6. N.P. Kurbatsky, *Forest fire science issues* (ILiD SO RAN, Krasnoyarsk, 1970)
7. A.V. Volokitina, M.A. Sofronov, *Vegetation fuel classification and mapping* (SO RAN, Novosibirsk, 2002)
8. M.A. Sofronov, I.G. Goldammer, A.V. Volokitina, T.M. Sofronova, *Wildland fire danger* (IL SO RAN, Krasnoyarsk, 2005)
9. A.V. Volokitina, T.M. Sofronova, M.A. Korets, *Improvement of fire danger rating in the forest* (IL SO RAN, Krasnoyarsk, 2018)
10. A.V. Volokitina, M.A. Sofronov, M.A. Korets, T.M. Sofronova, I.A. Mikhailova, *Forest fire behaviour prediction* (IL SO RAN, Krasnoyarsk, 2010)
11. M.A. Korets, A.V. Volokitina, *Software program for pyrological description of forest inventory plots* (State patent № 2014660252, 2014)
12. M.A. Korets, A.V. Volokitina, *Software program for surface fire spread prediction* (State patent № 2015661771, 2015)

# Comprehensive assessment of a flora and a productivity of green planting in the city of Abakan

*Elena* Yu. Zhukova*, *Elena* G. Lagunova

Khakas State University, 655000 Abakan, the Russian Federation

**Abstract.** The article provides information about the level of species diversity, biological and ecological composition, the dynamic of the vegetation productivity of a green planting in the city of Abakan (the steppe zone). The inventory of the flora revealed 288 species of higher vascular plants from 59 families and 183 genera. The basis of flora is the angiosperms, including 57 families (96.6 %), 179 genera (97.8 %), 282 species (97.9 %). The leading positions belong to the families Asteraceae (11.8 %), Poaceae and Rosaceae (10.8 %), Fabaceae (7.3 %). The most important genera in the flora were boreal ones – *Carex*, *Artemisia* (2.8 %), *Potentilla* (1.7 %). The ecological analysis showed the predominance of mesophytes (45.5 %). Perennial plants with clustered fleshy roots and short-rhizomatous ones (21.5 %) were a main biomorph. The analysis of the primary production of urban vegetation according to Terra MODIS data from 2000 to 2018 was carried out. The highest seasonal growth rates observed in 2001 and 2007, and the lowest ones were in 2004 and 2008. The average annual growth rates were 115±9 % for May, 107±5 % for June, 107±5 % for July, 119±14 % for July, and 120±11 % for August. The positive trend in the productivity of urban vegetation was established using the series criteria.

The city of Abakan (Khakasia) is located in the central part of the Minusinsk Basin (Uybat steppe) and covers an area of 112.87 $km^2$. The city is located in an area with frequent inversions, which makes it difficult to disperse pollutants. The main types of green spaces in the city are parks, squares, street plantations and public lands, which occupy a third of the area. The collection of higher vascular plants was carried out by route method from May to early September from 2012 to 2017 to cover the periods of the flowering of different systematic groups. In the laboratory, the binary name of the collected plants was determined by the Flora of Siberia T. 1 – 14 (1987 – 2003) [1].

The source satellite data represented a common 8-day composite of growth primary production (GPP) [2]. For retrieving Terra MODIS data we used the site https://lpdaacsvc.cr.usgs.gov/appeears. In the paper a discrete one-dimensional time series for equidistant 8-day moments was considered. To characterize the series, we used the growth rate – the ratio of the subsequent level to the previous one (%).

Several factors influence the formation of time series values:

$$x(t) = T(t) + S(t) + C(t) + N(t), t = \overline{1, P}, \quad (1)$$

* Corresponding author: biosara@mail.ru

where T(t) – trend component; S(t) – seasonal component; C(t) – cyclic and N(t) – irregular components [3].

The presence or absence of a nonrandom (and time-dependent t) component in the decomposition of time series (1) was determined according to the series criterion based on the median [3]. The value of the series is compared with the sample median, and if x (t) > , then for the corresponding observation, the member of the sequence forming the series takes the sign "+", if x (t) < , then the sign "-". The hypothesis of a randomness of a series is not rejected, if the inequalities are carried out (for 5% level of a significance): (2)

$$\tau_{\max}(n) < [1.43\ln(n+1)]$$
$$v(n) > \left[\frac{1}{2}(n+2-1.96\sqrt{n-1}\right], \qquad (2)$$

where n – the length of the time series; v(n) – the number of series; τ max (n) – the number of consecutive pluses or minuses in the longest series.

The module «Time Series/Forecasting» of the program «Statistica» was used to perform time series decomposition according to Census 1 method.

There were 288 species of higher vascular plants belonging to 59 families and 183 genera in the flora of parks and squares of Abakan. The basis of flora were angiosperms, including 57 families, which were (96.6 %), 179 genera (97.8 %), 282 species (97.9 %). Vascular cryptogamous plants comprised the Equisetaceae family (1.7 %), *Equisetum* genus (0.6 %) and 2 species (0.7 %): *Equisetum arvense* and *E. pratense*. The composition of gymnospermous plants included Pinaceae family (1.7 %), 3 genera – *Larix*, *Picea*, *Pinus* (1.6 %) and 4 species (1.4 %): *Larix sibirica* Ledeb., *Picea obovata* Ledeb., *Picea pungens* Engelm., *Pinus sylvestris* L.

Ten leading families comprise 62.2% of the species of the flora of the parks and gardens of the city of Abakan. The leading positions belong to the largest families of Asteraceae (34 species, 11.8% of the total number of species), Poaceae and Rosaceae (31 species or 10.8 %). Asteraceae and Poaceae families are the most common in the flora of Khakasia. The high position of Rosaceae in the spectrum was explained by the presence of numerous tree-shrubed adventive species (13 species) used in the greening. In the Rosaceae family *Pyrus ussuriensis* Maxim., *Rosa majalis* Herrm.,*Padus avium* Mill., *Spiraea chamaedryfolia* L., *Sorbus sibirica* Hedl., *Sorbaria sorbifolia* (L.) A. Braun, *Potentilla anserina, Geum aleppicum* Jacq. are the most common species. Representatives of the Fabaceae family (21 species, 7.3 %) are characterized by high adaptation to urbanization (*Trifolium pratense* L., *Amoria repens* (L.) C. Presl, *Medicago sativa* L., *Onobrychis arenaria* (Kit.) DC., *Vicia cracca* L.).

Among ten most common families are *Brassicaceae* (5.5%), *Ranunculaceae* (3.8%), *Lamiaceae* (3.5 %), *Cyperaceae* (3.1 %), *Boraginaceae* и *Caryophyllaceae* (2.8 %). Groups of plants with different attitudes to soil moisture were identified for the ecological analysis [4]. Most of the flora consists of mesophytes (131 species, 45.5 %). It is normal, since there are enough habitats with favorable water regime (meadows, drainage channels banks, etc.). A group of mesoxerophytes (75 species or 26.0 %), growing along roads, paths, steppe meadows, on the outskirts of green zones, was the second. Less significant was the group of xerophytes, which has 39 species (13.5%). There are some of hygrophytes in the city (22 species or 7.6 % of the total number of species). Mesohygrophytes (17 species or 6.0 %) and hydrophytes (4 species, 1.4%) were less common. Favorable conditions for the growth of hydrophytes are created in the coastal zone of numerous water sources.

It is possible to characterize habitat conditions based on the analysis of plant life forms according to I. G. Serebryakov (1962) [5]. The main biomorphological group is the group of perennial herbaceous plants (165 species, or 57.3% of the total number of species). The most numerous were perennials with clustered fleshy roots and short-rhizome ones, including 62 species (21.5 %). The long-rhizomatous plants are represented by 44 species (15.3 %). The composition of rod-rooted plants growing in steppe areas and steppe meadows, along the roadsides and in the areas of buildings included 31 species (10.8 %). Loose-bunchgrass (15 species, 5.2 %) and dense-bunchgrass (9 species, 3.1%) belonged mainly to the Poaceae family. Bulbous plants were presented by four species (1.4%) of Alliaceae (*Allium anisopodium* Ledeb., *A. ramosum* L. etc) and *Lilliaceae (Gagea fedtschenkoana* Pasch., *Tulipa heteropetata* Ledeb., *T. uniflora* (L.) Besserex Backer) families. There were many annual herbaceous plants (45 species, 15.6%) in the city. Many of them are weed species, which is associated with a strong anthropogenic factor. Biennials are represented by 25 species, or make up 8.7% of the total number of species. A significant part of shrubs (30 species, 10.4 %) belongs to *Rosaceae* family: *Sorbaria sorbifolia, Padus avium, Spiraea media* Schmidt etc. Among the trees (20 species, 7.0 %), *Ulmus pumila* L., *Betula pendula* Roth, species from the genus *Populus*, and species of the family *Pinaceae* are common. There are 2 species (1.6%) of dwarf subshrubs (*Artemisia frigida* Willd. and *Potentilla bifurca* L.), and 1 species of subshrubs (*Solanum kitagawae* Schonb.-Tem.). Perennial plants are more adapted to the inclement conditions than annual and biennial plants.

According to data from the Terra MODIS primary production, the growth rates for the time series from 2000 to 2018 were calculated (Fig. 1).

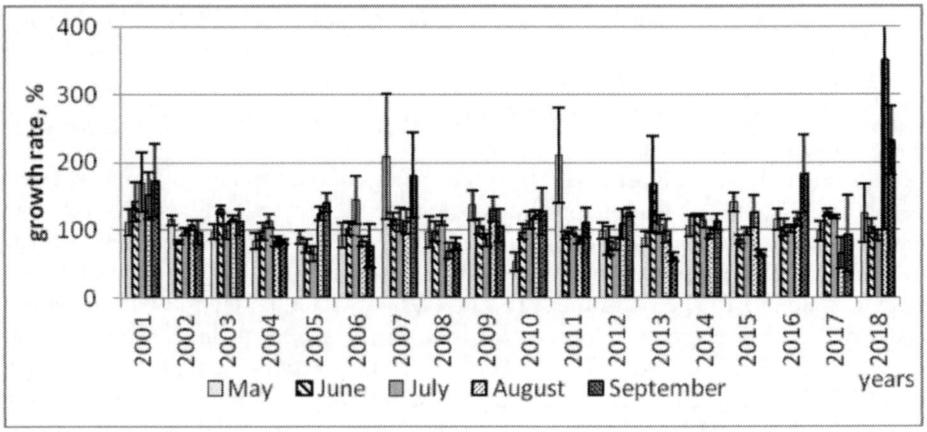


**Fig. 1.** Growth rates of the primary growth production of vegetation (Abakan, 2001-2018)

As can be seen in Figure 1, the maximum growth rates were noted in 2001 and 2007. Minimum values were found in 2004 and 2008. Intensive growth rates were observed not only in May (2007, 2009, 2011), but also in August (2001, 2005, 2007, 2012, 2016, 2018). It is typical of the steppe zone (the second peak of the growing season). Absolute maximum of growth rates for the entire period (350%) was recorded in 2018. It is associated with favorable weather conditions and the appearance of new green zones, planting of new rapidly growing trees and lawns. To establish the presence of a trend, the values ($n=398$, $\nu(n) = 45$, $\tau\ max = 14$) were substituted into the system (2). Both inequalities violated; therefore, hypothesis (2) rejected, i.e. in the original time series there was a non-random component. Since the components of the series depend on the level of productivity of the season, a multiplicative model of interaction was chosen.

As a result of seasonal decomposition, components involved in the formation of a time series using the Census 1 method (Fig. 2-3) are highlighted. The dynamics of the vegetation productivity of the city of Abakan indicated typical seasonal fluctuations overlapping the trend. The indicator was in the range of 90-115 %. The growth rate is determined by the phenology of woody deciduous vegetation in the arid zone.

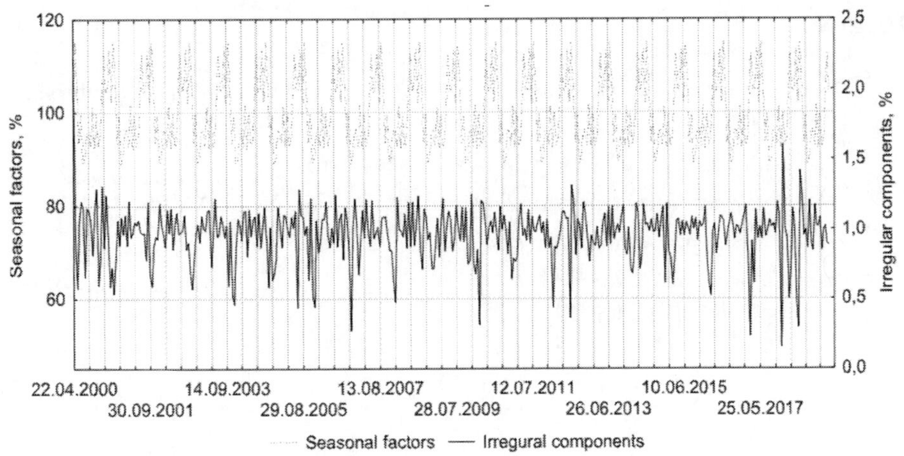


**Fig. 2.** Seasonal and irregular components of the primary production (Abakan, 2001-2018)

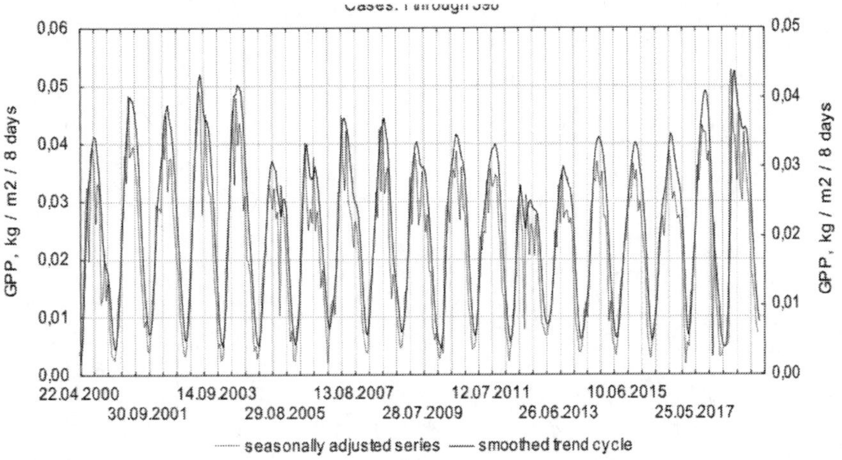


**Fig. 3.** Trend cycle of the primary production (Abakan, 2001-2018)

The contribution of the irregular component (change in the area of the green zones, pollution, damage) amounted to an average of about 1.00% (in the range of 0.02-1.50%), mainly a decrease in performance.

After seasonal adjustment of the series, one and less often two peaks in the development of vegetation were found (Fig. 3). The trend cycle showed maximum values (≥0.04 kg / m$^2$ / 8 days) of the growth productivity indicator in 2001-2004 and in 2017-2018. Minimum values (≤0.03 kg / m$^2$ / 8 days) were noted for 2005 and 2012. These periods coincided with severe drought. Satellite indicators of productivity correlated with the sum of positive temperatures ($R^2 = 0.603$) and the hydrothermal coefficient ($R^2 = 0.727$) in the study area.
The trend were on average 0.01 kg / m$^2$ / 8 days lower, compared to the seasonally adjusted graph. In a time series of traced wave-like cycle, the length of which 5-6 years and the positive trend were observed in 2017-2018.

## Acknowledgments

AppEEARS (https://lpdaacsvc.cr.usgs.gov/appeears)

## References

1. *Flora of Siberia*: in 14 volumes / edited by L.I. Malyshev (Novosibirsk, Nauka, 1987–2003)
2. S. Running, Q. Mu, M. Zhao, MOD17A2H (2015).
3. S.A. Aivazyan, *Fundamentals of econometrics* (YUNITI-DANA, 2001)
4. T.K. Goryshina, *The Ecology of plants* (Higher school, 1979)
5. I.G. Serebryakov, *Ecological morphology of plants. Life forms of angiosperms and conifers* (Higher school, 1962)

# Cryopetrophytic communities of the Altai-Sayan mountain system

*Evgeny* Zibzeev*, *Natalia* Igai

Central Siberian Botanical Garden, Siberian Branch of the Russian Academy of Sciences, 630090, 101 Zolotodolinskaya str., Novosibirsk, Russia

**Abstract.** Floristic classification of the cryopetrophytic alpine communities of the Altai-Sayan mountain system have been performed. The ***Rhodioletea quadrifidae*** Hilbig 2000 class of the Altai-Sayan mountain system includes two order, one alliance and six associations. We were described new order (***Valerianetalia petrophyllae*** order nova prov.) and three associations (***Ranunculo akkemensis–Valerianetum petrophilae*** ass. nova prov., ***Mesostemmo martjanovii–Leiosporetum exscapae*** ass. nova prov., ***Potentillo biflorae–Sibbaldietum tetrandrae*** ass. nova prov.) These associations include communities with a predominance of Asian alpine and arctalpine petrophytes.

## 1 Introduction

Cryopetrophytic communities of the Altai-Sayan mountain system as those of the Asian highlands exhibit a unique type of vegetation confined to cold rock habitats. In the South-East Altai and central part of the West Sayan, they represent a type of vegetation characterized the belt [1, 2]. Despite their wide distribution, these communities are the infrequently studied.

In terms of the Russian geobotanical school, the communities formed on scree slopes are rarely considered an independent assemblage. Most often they are referred to "scree communities" or "rock and scree vegetation" [3–5]. We do not agree with other researchers [4] that "scree slopes are not distinct in the originality of their flora." Our data demonstrate that these communities are characterized by high heterogeneity in species composition, and this is due to numerous factors: the feature of the underlying rocks, the mobility of the substrate, the nature and structure of scree slopes, chemical composition, the rate of silt accumulation, steepness of slopes, slope exposure, etc. Various combinations of these factors lead to formation of high coenosis diversity. R. V. Kamelin named these cryopetrophytic communities "cryopetrophyton oreoasiaticum" and conside them as different vegetation types [1].

The system of higher units of floristical classification of the cryopetrophytic communities of the Altai-Sayan mountain system was proposed by V. Hilbig [6, 7, 8]. Based on the geobotanical data collected in the highlands of northern Mongolia, he identified the class ***Rhodioletea quadrifidae*** Hilbig 2000 with two associations: ***Saxifrago***

* Corresponding author: egzibzeev@gmail.com

***setigerae-Rhodioletum quadrifidae*** Hilbig (1987) 1990 and ***Oxygraphio glacialis-Potentilletum biflorae*** Hilbig (1987) 1990. We found these communities on Katun Range of the Central Altai [9] and Sailig-Hem-Taiga Ridge of the Western Sayan [2]. We described new association of ***Saxifrago oppositifoliae-Rhodioletum quadrifidae*** Zibzeev 2013 on the territory of the Western Sayan [2].

An integrated classification system of the ***Rhodioletea quadrifidae*** class does not developed yet and there is no consensus of the authors about higher units of cryopetrophytic alpine communities.

This paper presents ecological characterization of the communities of the class ***Rhodioletea quadrifidae*** the Altai-Sayan mountain system and their coenotic diversity .

## 2 Study area

The highest mountainous area of the Altai-Sayan mountain system at altitudes of 1700-2900 m is dominated by ancient table-lands. The highest ridges (altitudes 2500-4500 m) havesteep slopes on metamorphic rocks, mainly base-rich chloride slates, with igneous rocks and limestones occupying small areas. Due to its location in the centre of Eurasia, the macroclimate of the Altai-Sayan mountain system is strongly continental. Local climate, however, is considerably modified by its mountainous topography.

## 3 Material and methods

The materials for this paper were 45 relevés performed using a standard technique when studying the alpine vegetation of the Altai-Sayan mountain system in 2012-2017. The Braun-Blanquet approach was used to classify the plant communities. The relevés were processed with the automatic classification technique using JUICE 7.0 software package. The nomenclature of the syntaxonomical units was done in accordance with the "International code of phytosociological nomenclature" [10]. Latin names of species and nomenclature follow The International Plant Names Index (IPNI) (http://ipni.org/).

## 4 Results and discussion

The class ***Rhodioletea quadrifidae*** Hilbig 2000 includes alpine petrophytic communities of the mountain systems of Siberia, northern Mongolia and East Kazakhstan, growing on eluvial scree slopes. These communities are characterized by a high heterogeneity in species composition due to factors such as altitude, slope exposure, slope steepness, nature, size and composition of rock screes. Such factors as the rate of silt accumulation and soil formation significantly affect the qualitative composition of these communities. Currently, the class is represented by alliance ***Rhodilion quadrifidae*** Hilbig 2000 and tow order ***Rhodioletalia quadrifidae*** Hilbig 2000 and ***Valerianetalia petrophyllae*** order nov. prov.

Class ***Rhodioletea quadrifidae*** Hilbig 2000
Ord. ***Rhodioletalia quadrifidae*** Hilbig 2000
All. ***Rhodilion quadrifidae*** Hilbig 2000
Ass. ***Saxifrago oppositifoliae–Rhodioletum quadrifidae*** Zibzeev 2013
Ass. ***Saxifrago setigerae–Rhodioletum quadrifidae*** Hilbig (1987) 1990
Ass. ***Oxygraphio glacialis–Potentilletum biflorae*** Hilbig (1987) 1990
Ass. ***Potentillo biflorae–Sibbaldietum tetrandrae*** ass. nova prov.
Ord. ***Valerianetalia petrophyllae*** order nova prov.
All. ?

Ass. ***Ranunculo akkemensis–Valerianetum petrophilae*** ass. nova prov.
Ass. ***Mesostemmo martjanovii–Leiosporetum exscapae*** ass. nova prov.

Based on the analysis of coenoflora of all associations, we propose the following composition of the diagnostic group of the class ***Rhodioletea quadrifidae***: *Cardamine bellidifolia* L., *Carex rupestris* Turcz. ex Ledeb., *Cerastium pusillum* Ser., *Gastrolychnis apetala* (L.) Tolm. et Kozhanch., *Lagotis integrifolia* (Willd.) Schischk., *Lloydia serotine* (L.) Salisb. ex Rchb., *Luzula confuse* Lindeb., *Minuartia arctica* (Steven) Graebn., *M. verna* (L.) Heirn, *Oxygraphis glacialis* (Fisch.) Bunge, *Papaver pseudocanescens* Popov, *Rhodiola quadrifida* Fisch. et C.A. Mey., *Saxifraga cernua* L., *S. oppositifolia* L., *Smelowskia alba* B. Fedtsch., *Smelowskia calycina* (Stephan ex Willd.) C.A. Mey., *Thalictrum alpinum* L. [6].

The order ***Rhodioletalia quadrifidae*** unites the communities developing on stabilized alpine scree. It includes one alliance and four associations.

Ass ***Saxifrago oppositifoliae–Rhodioletum quadrifidae*** Zibzeev 2013.

Diagnostic species: *Campanula dasyantha* M. Bieb., *Carex tristis* M.Bieb. subsp. *stenocarpa* (Turcz. ex V.I.Krecz.) T.V.Egorova, *Cerastium lithospermifolium* Fisch., *Hierochloë alpina* Roem. & Schult., *Paraquilegia microphylla* J.R.Drumm. & Hutch., *Salix berberifolia* Pall., *Saussurea foliosa* Ledeb., *S. schanginiana* (Wydler) Fisch. ex Herder, *Saxifraga bronchialis* L., *S. hirculus* L., *S. melaleuca* Fisch. ex Spreng., and *S. nelsoniana* D. Don.

The association ***Saxifrago oppositifoliae–Rhodioletum quadrifidae*** includes the communities with a predominance of alpine and arctic-alpine petrophytes, most of which are facultative and obligate calcicoles. This association occurs not only in the West Sayan Mts., but also in the west and central parts of the East Sayan [2], and North-West and Central Altai [1, 9]. Typically these cenoses do not occupy large areas. They occur on the eluvial scree slopes of relatively flat tops of ridges and extending slopes (up to 25°) at altitudes of 2300–2570 m above sea level in moderately humid sites.

Ass. ***Saxifrago–Rhodioletum quadrifidae*** Hilbig (1987) 1990

Diagnostic species: *Rhodiola quadrifida* (Pall.) Fisch. & C.A. Mey., *Lagotis integrifolia (Willd.) Schischk.*, *Eremogone meyeri* (Fenzl) Ikonn., *Potentilla sericea* L., *Eritrichium pauciflorum* (Ledeb.) DC., *Saxifraga setigera* Pursh., *S. cernua* L., *S. hirculus* L., *Eutrema edwardsii* R. Br., *Draba fladnizensis* Wulfen.

These communities occur in the upper part of the alpine belt on shallow stony soils and rock debris. They were found in the highlands of the South Tuva, Hangai and the Mongolian Altai at altitudes from 2600 to 3600 m. *Rhodiola quadrifida* (Pall.) Fisch. & C.A. Mey., *Lagotis integrifolia (Willd.) Schischk.*, *Eremogone meyeri* (Fenzl) Ikonn., *Potentilla sericea* L., and *Eritrichium pauciflorum* (Ledeb.) DC. predominate.

Ass. ***Oxygraphio glacialis–Potentilletum biflorae*** Hilbig (1987) 1990

Diagnostic species: *Oxygraphis glacialis* (Fisch.) Bunge, *Potentilla biflora* Willd. ex Schltdl., *Saxifraga oppositifolia* L., *Poa alpine* L., *Cerastium pusillum* Ser.

These high-mountain communities inhabit the alpine belt of Northern Mongolia and the southern mountain ranges of the Altai-Sayan mountain system (2500-3950 m). The vegetation cover is from 30 to 60 %. In a relevé 5-16 species are recorded. In addition to diagnostic species, *Festuca altaica* Trin., *Oxytropis martjanovii* Krylov, *Polygonum viviparum* L., *Potentilla crebridens* Juz., *Saxifraga cernua* L., *Smelowskia alba* (Pall.) Regel, *Stellaria petraea* Bunge have high constancy.

Ass. ***Potentillo biflorae–Sibbaldietum tetrandrae*** ass. nova prov.

Diagnostic species: *Sibbaldia tetrandra* Bunge, *Rhodiola algida* (Ledeb.) Fisch. & C.A. Mey., *Cerastium lithospermifolium* Fisch., *Lloydia serotina* (L.) Rchb..

This association is widely distributed in the South-East Altai (Kuray, South-Chuyskiy, Chikhacheva). Patches of this association occur from 1650 to 2900 m. Soil layer is extremely shallow or completely lacking. The vegetation cover is 40 to 75 %. *Sibbaldia tetrandra* is the absolute dominant of the community; other species (*Cerastium lithospermifolium* Fisch., *Gastrolychnis apetala* (L.) Tolm. & Kozhanch., *Poa alpina* L., *Potentilla nivea* L., *Rhodiola quadrifida* (Pall.) Fisch. & C.A. Mey., *Smelowskia calycina* (Stephan) C.A. Mey., *Taraxacum glabrum* DC.) have a high occurrence but low cover.

On the territory of South-East Altai (Kuraisky, Chikhachev and South-Chuisky ranges), we described cryopetrophytic communities are formed on unstable alpine scree. Soil layer is completely lacking. We propose to include these communities in the new order ***Valerianetalia petrophyllae*** order nov. prov. The order includes two associations ***Ranunculo akkemensis–Valerianetum petrophilae*** and ***Mesostemmo martjanovii–Leiosporetum exscapae***.

Ass. ***Ranunculo akkemensis–Valerianetum petrophilae*** ass. nova prov.

Diagnostic species: *Valeriana petrophila* Bunge, *Smelowskia calycina* (Stephan) C.A. Mey., *Ranunculus akkemensis* Polozhij & Revjakina.

The association includes communities on unstable alpine scree. Patches of the association are rather poor in species. In a relevé, there are 5-10 species. Cover of ***Ranunculo akkemensis–Valerianetum petrophilae*** is 10-20 %. High constancy species of this community include *Crepis nana* Richardson, *Eremogone formosa* (Fisch. ex Ser.) Fenzl, *Lagotis integrifolia* (Willd.) Schischk., *Minuartia arctica* (Steven ex Ser.) Graebn., *Ranunculus altaicus* Laxm..

Ass. ***Mesostemmo martjanovii–Leiosporetum exscapae*** ass. nova prov.

Diagnostic species: *Leiospora exscapa* (C.A. Mey.) F. Dvorak, *Mesostemma martjanovii* (Krylov) Ikonn.

These are rare communities dominated by *Leiospora exscapa* (C.A. Mey.) F. Dvorak and *Mesostemma martjanovii* (Krylov) Ikonn. These species are listed in the Red Book of Russia. This association is very similar with the community ***Ranunculo akkemensis–Valerianetum petrophilae*** association by structure and ecology. In a relevé, there are 5-12 species. Its vegetation cover varies from 10 to 30 %. Several species have high constancy: *Allium pumilum* Vved., *Dracocephalum bungeanum* Schischk. & Serg., *Papaver pseudocanescens* Popov, *Pulsatilla campanella* (Fisch. ex Regel et Tiling) Krylov., *Thymus narymensis* Serg. etc. are.

Thus, alpine vegetation on the screen and stone debris were ascribed to the class ***Rhodioletea quadrifidae***. The class includes two orders, one alliance and six associations. Cryopetrophytic communities of the Altai-Sayan mountain system possess a group of regional Asian cryopetrophytic alpine species. It is cryo-petrophytes species. The order of ***Rhodioletalia quadrifidae*** and ***Valerianetalia petrophyllae*** differ from each other not only by species composition, but also by habitat features (the presence of soil horizon and substrate mobility).

Acknowledgments: The reported study was funded by RFBR, project number 17-04-00076, State task of the Central Siberian Botanical Garden SB RAS (№ registration AAAA-A17-117012610052-2).

## References

1. R.V. Kamelin, *Brief characteristic of natural conditions and vegetation of Altai mountain country. Flora Altaya*, **1** (Barnaul: AZBUKA, (2005)
2. E. G. Zibzeev, Vestnik NSU: biology, clinical medicine, **11(1)**, 92 (2013)
3. L. I. Malyshev, *Alpine flora of the Eastern Sayan* (Moscow-Leningrad: Science, 1965)

4. I. M. Krasnoborov, *Vegetation of the right Bank Yenisei* (Novosibirsk: Science, 1971)
5. I. M. Krasnoborov, *Flora alpina montium salanesium occidentalium* (Novosibirsk: Science, 1976).
6. W. Hilbig, *The vegetation of communities of Mongolia* (Amsterdam: SPB Academic Publishing, 1996).
7. W. Hilbig, Feddes Repertorium **111**, 75 (2000)
8. W. Hilbig, *Pflanzengesellschaften der Mongolischen Volksrepublik* (Diss. B. Ms., Halle, 1987)
9. A. Yu. Korolyuk, *Flora and vegetation of the Katun reserve* (Novosibirsk, 2001)
10. V. Westhoff, E. van der. Maarel, Veg. Sci. **5**, 617 (1973)

# Extrazonal steppes of forest belt on eastern macroslope of the Urals

*Natalya* Zolotareva[1,*], *Andrey* Korolyuk[2]

[1]Institute of Plant and Animal Ecology UB RAS, 620144, 8 Marta Str., 202, Ekaterinburg, Russia
[2]Central Siberian botanical garden SB RAS, 630090, Zolotodolinskaya Str., 101, Novosibirsk, Russia

**Abstract.** Extrazonal steppes of forest belt on eastern macroslope of the Middle and South Urals have small coenotic diversity. The most part of studied communities are petrophytic steppes on outcrops, which determine regional features of plant cover and provide habitats to rare, endemic and relict plant species. Petrophytic steppes correspond to order ***Helictotricho-Stipetalia***, meadow steppes and xeric meadows, shrub thickets correspond to order ***Brachypodietalia pinnati*** (class ***Festuco-Brometea***).

Extrazonal steppe is characteristic vegetation element of forest belt of the Middle and South Urals. In the Middle Urals the steppes occur on basic and ultrabasic outcrops till the southern boundary of middle taiga. In boreal zone of the South Urals the steppes are found mainly on the hyperbasites of the eastern mountain ranges. The main part of studied communities are petrophytic steppes on outcrops and a little part are meadow steppes and xeric meadows on gentle slopes. The information about steppes of forest belt of the Urals is fragmentary [1, 2, 3]. The aim of our research was to investigate the diversity of extrazonal steppe communities of forest belt of the Urals on the territory of Sverdlovsk and Chelyabinsk Regions. The dataset includes 595 relevés collected in the forest belt of the Middle and South Urals. All relevés were classified using JUICE 7.0. The dataset presents 9 types of communities which differ in species composition and set of dominants; activity of species was calculated for each type of community. The first four clusters unify meadow steppes and xeric meadows, shrub thickets, the remaining clusters are petrophytic steppes.

Clusters:

1. Meadow steppes on deep soils of lower parts of slopes (the Middle Urals). Species of meadow steppes: *Filipendula vulgaris, Fragaria viridis, Poa angustifolia, Stipa pennata*. Meadow steppes are differentiated by species of dry meadows and thermophytic forest edges (*Astragalus danicus, Centaurea scabiosa, Galium album, Gentiana cruciata, Knautia arvensis, Leucanthemum vulgare, Plantago media, Potentilla argentea, Trifolium pratense*), creating the core of communities. Cluster includes communities of ass. ***Leucanthemo vulgaris–Stipetum pennatae*** Bayanov in Yamalov et al. 2013 [4].

2. Meadow steppes on shallow soils of stony slopes surrounded by forest (the Middle Urals). They are differentiated by species of forest edges (*Brachypodium pinnatum, Pinus sylvestris, Rubus saxatilis*) and petrophytic steppes (*Centaurea sibirica, Echinops crispus*). *Stipa pennata* and *Helictotrichon desertorum* usually dominate, sometimes *Calamagrostis*

---

* Corresponding author: nvp@ipae.uran.ru

*arundinacea, C. epigeios, Centaurea sibirica* and *Brachypodium pinnatum* take their place. Cluster includes communities of ass. ***Pulsatillo uralensis-Helictotrichetum desertorum*** Teptina et al. 2018 (subass. ***P.u.-H.d. calamagrostietosum arundinaceae*** and subass. ***P. u.-H. d. calamagrostietosum epigeii***,var. ***Stipa dasyphylla***).

3. Meadow steppes on shallow soils of middle and lower parts of stony slopes (the South Urals). They are differentiated by species of forest edges (*Hylotelephium triphyllum*), meadow steppes (*Artemisia latifolia, A. sericea, Cerastium arvense, Galatella angustissima, Helictotrichon schellianum*) and petrophytic steppes (*Eremogone micradenia, Silene repens, Thalictrum foetidum*). Species of meadow steppes (*Artemisia sericea, Fragaria viridis, Stipa pennata, Helictotrichon desertorum*) and species of forest edges *Calamagrostis epigeios* dominate.

4. Shrub thickets (the Middle and South Urals). Species of forest edges (*Aconogonon alpinum, Campanula bononiensis, Nepeta pannonica, Vicia tenuifolia*) differentiate shrub thickets; *Cerasus fruticosa, Rosa majalis, Spiraea crenata* have high activity. *Spiraea crenata* is the key dominant, sometimes *Cerasus fruticosa* accompanies it. Thickets of steppe shrubs occupy a large area in the forest belt of the South Urals, in the Middle Urals they make small patches.

5. Petrophytic variants of meadow steppes (the Middle Urals).These widespread communities do not have any differential species. They occur on slope outcrops surrounded by forest. The species of petrophytic steppes (*Centaurea sibirica, Echinops crispus, Helictotrichon desertorum, Onosma simplicissima*), typical steppes (*Festuca valesiaca*) and meadow steppes (*Stipa pennata*) have high activity in these communities. They belong to ass. ***Stipo pennatae–Centauretum sibiricae*** Yamalov et al. 2013 nom. invalid., ass. ***Pulsatillo uralensis-Helictotrichetum desertorum,*** and community ***Thymus uralensis-Dianthus acicularis*** [3].

6. Petrophytic variants of meadow steppes (the South Urals). These communities occupy outcrops on the upper and middle parts of slopes. High activity of petrophytes (*Clausia aprica, Lychnis sibirica, Scorzonera glabra*) differ them from the communities of previous cluster: they are diagnostic species of ass. ***Stipo pennatae–Centauretum sibiricae*** in this communities.

7. Xerophytic outcrop communities (the Middle Urals). Petrophytes *Alyssum obovatum* and *Vincetoxicum albowianum* have high activity. Diagnostic species of class ***Festuco-Brometea*** Br.-Bl. et Tx. ex Soó 1947 (*Festuca valesiaca, Galium verum, Phleum phleoides, Veronica spicata*) are also active.

8. Petrophytic steppes dominated by *Stipa capillata* and *Festuca valesiaca* (the Middle and South Urals). The communities of steep rocky slopes are differentiated by species of forest edges (*Potentilla argentea*), meadow steppes (*Achillea nobilis, Thymus marschallianus*) and typical steppes (*Stipa capillata, Verbascum phoeniceum*). The high activity of petrophytes (*Artemisia frigida, Dianthus acicularis, Echinops crispus, Koeleria cristata, Thymus punctulosus, Veronica spicata*) is a distinctive feature of these communities.

9. Petrophytic communities of riverbank outcrops (the Middle Urals). Petrophytes (*Schiverecki a hyperborea, Minuartia kraschenіnnikovii*) and species of steppes and ruderal habitates (*Dracocephalum thymiflorum*) characterize them. Petrophytes (*Elytrigia reflexiaristata, Aster alpinus, Carex pediformis*) and weed *Arenaria serpyllifolia* have high activity. Relevés include diagnostic species of class ***Festuco-Brometea*** (*Campanula sibirica, Festuca valesiaca, Veronica spicata*) and order ***Helictotricho-Stipetalia*** Toman 1969 (*Helictotrichon desertorum, Potentilla humifusa*).

Cluster analysis revealed 2 group of communities (Fig.). Group A brings together meadow steppes, xeric meadows, and shrub thickets; it corresponds to order ***Brachypodietalia pinnati*** widespread in Europe [5]. The group B includes petrophytic steppes, and corresponds to order ***Helictotricho-Stipetalia*** (typical steppes and more xeric variants of meadow steppes of Urals, Northern Kazakhstan and West Siberia [6].

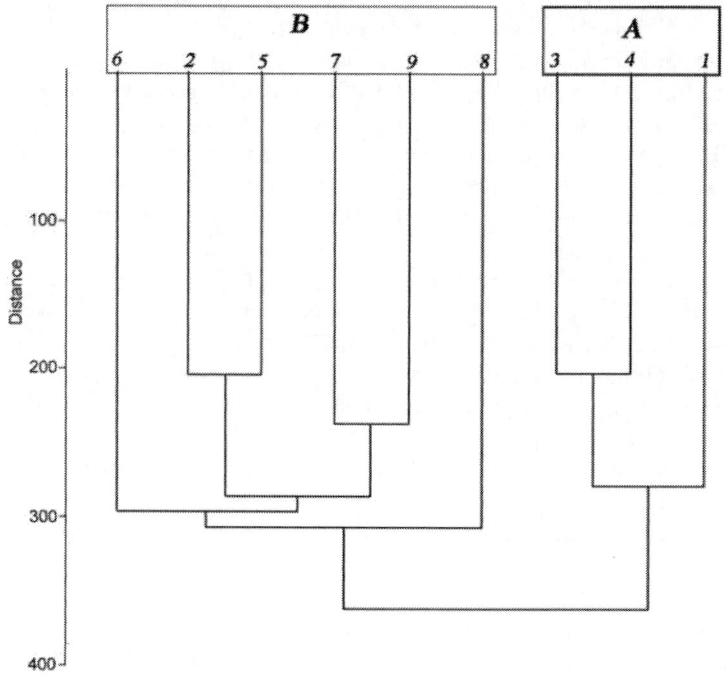


**Fig.** Cluster analysis (algorithm UPGMA, similarity index Euclidean distance). A – the order ***Brachypodietalia pinnati***, B – the order ***Helictotricho-Stipetalia***, 1-9 – correspond to communities in the text.

The studied vegetation has low coenotic diversity. The most part of meadow steppes and xeric meadows communities correspond to ass. ***Leucanthemo vulgaris–Stipetum pennatae***. Petrophytic steppes determine regional features of plant cover and provide habitats to rare, endemic and relict plant species. The main part of petrophytic communities belongs to ass. ***Stipo pennatae–Centauretum sibiricae***; some new syntaxa can be revealed. Many petrophytic communities of Sverdlovsk Region are included in nature monuments or are promising for organization of protected areas [7]. Petrophytic steppes of the boreal zone in the South Urals are protected only in Ilmensky Reserve whereas petrophytic communities of Vishnevye and Potaniny, Sugomak, Egozinskaya mountains are rich in rare plant species, one of them is local endemic *Oxytropis ponomarjevii*. Protected areas should be organized to preserve steppe vegetation in these habitats [2, 8].

This research was funded by RFBR according to the research project № 17-04-00276, as well as within the framework of the state task of the Institute of Plant and Animal Ecology.

## References

1. V.B. Sochava, Sov. Bot., **13**, № 3 (1945)
2. P.L. Gorchakovskii, N.V. Zolotareva, Russian Journal of Ecology, **37**, 6 (2006)
3. A.Y. Teptina, M.V. Lebedeva, S.M. Yamalov, Veg. Russ, 33 (2018)

4. S.M. Yamalov, A.V. Bayanov, A.A. Muldashev, E.A. Averinova, Veg. Russ, 22 (2011)
5. W. Willner et al., Appl. Veg. Sci., **20** (1) (2017)
6. A.Yu. Korolyuk, Veg. Russ., 30 (2017)
7. N.V. Zolotareva, E.N. Podgaevskaya, Izvestia RAS SamSC, **14**, 1 (4) (2012)
8. N.V. Zolotareva, Mountain Ecosystems and Their Components, **2** (2007)

# The Lichens of the Mountain-Taiga Belt of the Ergaki Nature Park

*Olga* A. Zyryanova*

Khakas State University named after N.F. Katanov, 90 Lenin St., 655000, Abakan, Russia

**Abstract.** The article presents the data on the species composition of lichens in the mountain-taiga belt of the Ergaki Nature Park. The taxonomic list of lichens is represented by 288 taxa of a species rank belonging to 89 genera and 39 families. The largest number of species is characterized by the crustose life form. Mesophytes and epiphytes are predominant in the terms of ecological substrate.

## 1. Introduction

The natural park of regional importance – Ergaki located in the Western Sayan occupies several districts of Krasnoyarsk Territory at once, specifically Karatuzsky, Ermakovsky and Kuraginsky districts. Due to its length it covers various high-altitude mountain belts.
The studies on the species composition of lichens were conducted in the vicinity of the Tushkanchik cordon, which is located according to the paper of M.V. Bocharnikov in the mountain-taiga belt of the Western Sayan [1].

## 2. Material and methods

The materials for the study include the herbarium specimens of lichens collected by the authors in the vicinity of the Tushkanchik cordon with the use of route method. The material collection and herbarization was performed using the common method described in the second edition of Identification Guide to Lichens of the USSR "Morphology, Systematics and Geographical Distribution" [2]. Around 360 herbarium envelopes were collected and then processed in the Laboratory of the Khakas State University named after N.F. Katanov with the use of the microscope Mikromed-1 and binocular loupe MBS-2 with 10% KOH solution. In order to identify the materials "Identification Guide to Lichens of the USSR" [3-6] and "Identification Guide to Lichens of Russia" [7-11] were used. Some types were specified in work of R. Santesson et al. [12]. The life forms and ecological groups were identified on the basis of the personal observations of the authors and according to the literary sources [13].

## 3. Results and discussion

* Corresponding author: o_a_zyryanova@mail.ru

As a result of the studies it was established that the lichen flora of the mountain-taiga belt of the Ergaki Nature Park in the vicinity of the Tushkanchik cordon is represented by 288 species belonging to 89 genera and 39 families.

The average number of species in a family is 7.38. As shown by taxonomic analysis the lichen flora is represented by 10 dominant families (76.74 %), among which the highest level of diversity belongs to the family of Parmeliaceae Zenker (68 species, 23.61 %). The further families located in the descending order include Cladoniaceae Zenker (50 species; 17.36%), Lecanoraceae Körb (27 species, 9.38%), Rhizocarpaceae M. Choisy ex Hafellner (14 species, 4.86 %), Physciaceae Zahlbr and Stereocaulaceae Chevall. (12 species; 4.17 %), Hymeneliaceae Körb, Peltigeraceae Dumor and Umbilicariaceae Chevall. (10 species, 3.47 %), Pertusariaceae Körb.ex. Körb. (8 species, 2.78 %). Among the families not included in the dominant category there are Porpidiaceae Hertel et Hafellner ranked 11th and having 6 species, the family of Lecideaceae Chevall having 5 species, Bacidiaceae W. Watson, Nephromatacaea Wetm.ex J. David et D. Hawksw. and Roccellaceae Chevall. having 4 species, 6 families having 3 species each, 7 families having 2 species each and 11 families having 1 species each.

89 genera found in the lichen flora. The average number of species in a genus is 3.24. The level of species diversity exceeding average can be found in 18 genera uniting 191 species (66.32 %). The remaining 21 genera account for 97 species (33.68 %), among which 3 species are represented by 9 genera (9.38 %), 2 species by 12 genera (8.33 %) and 1 species by 46 genera (15.97 %).

The spectrum of life forms is characterized by the predominance of the crustose lichens (125 species, 43.40 %). This thallus cannot be separated from the substrate where it grows without damaging it. As can be seen, for example, in *Amandinea punctata* (Hoffm.) Coppins et Scheid., *Caloplaca holocarpa* (Hoffm. Ex Ach.) A. E. Wade, *Aspicilia laevata* (Ach.) Arnold and others. The fruticose lichens (89 species, 30.90 %) and foliose lichens (74 species, 25.69 %) are less represented. Among the fruticose lichens the most common are *Cladonia arbuscula* (Wallr.) Flot.ssp. mitis (Sandst.) Ruoss, *Cl. amaurocraea* (Flörke) Schaer., *Cl. stellaris* (Opiz) Pouzar et Vêzda, *Bryoria implexa* (Hoffm.) Brodo et D. Hawksw., *Usnea glabrescens* (Nyl. ex Vain.) Vain., *U. longissima* Ach., al eds.; while the foliose lichens are represented by *Hypogymnia vittata* (Ach.) Parrique, *H. physodes* (L.) Nyl., *Peltigera polydactylon* (Neck.) Hoffm. *Melanelia septentrionalis* (Lynge) Essl., *Parmelia omphalodes* (L.) Ach., *P. sulcata* Taylor and others.

While establishing ecological groups of lichens in the vicinity of the Tushkanchik cordon, we used the work by N.V. Sedelnikova [13-14]. Seven ecological groups were identified: mesophytes, xerophytes, xeromesophytes, xerocryophytes, cryophytes, psychrophytes and hygrophytes. The predominant role belongs to the mesophytic lichens (209 species, 72.57 %). These are lichens that can be found in places with fairly moderate humidity, usually located on the shaded rocks or at their base. These include the lichen species such as *Aspicilia lapponica* Hue, *Physcia phaea* (Tuck.) Thoms., *Squamarina cartilaginea* (Witt.) P. James and others. The cryophytes rank second in terms of the species number (which makes 25). They account for 8.68% of the total number of species. These are the species of the coldest and driest habitats of the high-mountain belt. For example, *Lecidea auriculata* Th. Fr., *Miriquidica leucophaea* (Flörke et Rabenh.) Hertel et Rambold and others. The xerophytes rank third (23 species, 8.09 %). This group includes mainly those lichens, which are abundant in the conditions of dry and warm climate regimes. The examples include such species as *Lecanora frustulosa* (Dicks.) Ach., *L. argopholis* (Ach.) Ach., *Protoparmeliopsis muralis* (Schreb.) M.Choisy and others.

Based on the available data on the confinedness to the substrate in the studied area, the epiphytes hold a leading position (117 species, 40.63 %). They include, for example, *Ramalina pollinaria* (Westr.) Ach., *Rinodina sophodes* (Ach.) A. Massal., *Physcia tenella*

(Scop.) DC. and others. The group of epilithophytes is the least presented group of lichens (93 species, 32.29 %). For instance, *Porpidia cinereoatra* (Ach.) Hertel et Knoph, *Physcia caesia* (Hoffm.) Fürnr., *Lasallia pensylvanica* (Hoffm.) Hano and others. The soil cover species are represented by the epigeal lichens (78 species, 27.08 %) mainly from the genera of Cladonia, Peltigera and Stereocaulon.

The geographical analysis of lichens of the studied area was carried out by the principle of zonality based on the works of N.V. Sedelnikova [13-14]. All types are distributed into 7 geographical together with 13 types of natural habitats distributed in accordance with the nature of latitudinal placement of species (tab. 1).

**Table. 1.** Distribution of lichens of a mountain-taiga belt of the Ergaki on geographical elements and types of areas

| Geographical elements | Types of areas, number of species | | | | | | | | | | | | | Total number |
|---|---|---|---|---|---|---|---|---|---|---|---|---|---|---|
| | 1 | 2 | 3 | 4 | 5 | 6 | 7 | 8 | 9 | 10 | 11 | 12 | 13 | |
| Arctalpine | – | 1 | 11 | 12 | 7 | 9 | – | – | – | – | – | – | – | 40 |
| Alpine | – | – | – | – | – | – | – | – | 1 | – | – | – | – | 1 |
| Hypoarctmontan | – | 2 | 14 | 17 | 9 | 3 | 1 | – | – | – | – | 1 | | 47 |
| Montan | – | 4 | 10 | 17 | 17 | 4 | – | 1 | – | – | – | 3 | 1 | 57 |
| Boreal | – | 4 | 23 | 21 | 38 | 14 | – | – | – | 1 | – | 1 | – | 102 |
| Nemoralis | 1 | 3 | 2 | 10 | 10 | – | – | – | – | – | – | – | – | 26 |
| Steppe | 1 | 2 | 1 | 5 | 5 | – | – | – | – | – | 1 | – | – | 15 |
| TOTAL: | 2 | 16 | 61 | 82 | 86 | 30 | 1 | 1 | 1 | 1 | 1 | 5 | 1 | 288 |

**Note.** Types of natural habitats: 1 – Asian, 2 – Euro-Asiatic, 3 – Euro-American, 4 – Holarctic, 5 – Pluriregional, 6 – Holarctic-Notarctic, 7 – Euro-African, 8 – Submediterranean, 9 – Central Asian, 10 – Asian-Caucasian, 11 – African-Asian, 12 – American-Asian, 13 - American-Asian-African-Australian.

The geographical analysis of lichens demonstrated the predominance of the boreal geographical element (102 species, 35.4 %) with the inclusion of montane (57 species, 19.8%) elements. Among the species of boreal element the most common are epiphytic lichens *Hypogymnia physodes* (L.) Nyl., *H. vittata* (Ach.) Parrique, *Usnea. glabrescens* (Nyl. ex Vain.) Vain., *U. longissima* Ach., *Bryoria fuscescens* (Gyeln.) Brodo et D. Hawksw., *Evernia mesomorpha* (Müll. Arg.) Du Rietz, and from montane - *Lecanora varia* (Hoffm.) Ach., *Lecidella elaeochroma* (Ach.) M. Choisy, *Buellia disciformis* (Fr.) Mudd.

Analyzing the type of habitat one can say that the pluriregional (86 species, 29.9 %) and holarctic (82 species, 28.5 %) lichens are predominant.

The studied area comprised 4 species of lichens included in the Red Book of the Krasnoyarsk Territory[15]: *Lobaria pulmonaria* (L.) Hoffm., *Lobaria scrobiculata* (Scop.) DC., *Nephromopsis laureri* (Kremp.) Kurok., *Usnea longissima* Ach.

## 4. Conclusion

Thus, the lichen flora of the studied territory is basically boreal, as evidenced by the high grade of the dominant families, the biomorphological and ecological features of the species and the geographical analysis.

## References

1. M.V. Bocharnikov, *The geography of Botanical diversity of the Western Sayan mountains*, Diss. (Moscow, 2012)

2. A.N. Oksner, *Identification Guide to Lichens of the USSR*, **2** (Science, Leningrad, 1974)
3. *Identification Guide to Lichens of the USSR*, **1** (Science, Leningrad, 1971)
4. *Identification Guide to Lichens of the USSR*, **3** (Science, Leningrad, 1975)
5. *Identification Guide to Lichens of the USSR*, **4** (Science, Leningrad, 1977)
6. *Identification Guide to Lichens of the USSR*, **5** (Science, Leningrad, 1978)
7. *Identification Guide to Lichens of Russia*, **6** (Science, St. Petersburg, 1996)
8. *Identification Guide to Lichens of Russia*, **7** (Science, St. Petersburg, 1998)
9. *Identification Guide to Lichens of Russia*, **8** (Science, St. Petersburg, 2003)
10. *Identification Guide to Lichens of Russia*, **9** (Science, St. Petersburg, 2004)
11. *Identification Guide to Lichens of Russia*, **10** (Science, St. Petersburg, 2008)
12. O.E. Eriksson, D.L. Hawksworth, *Outline of the Ascomycetes* (Wallingford, UK, 1998)
13. N.V. Sedelnikova, *Lichens of Altai and Kuznetsk highlands: Summary of flora* (Science, Novosibirsk, 1990)
14. N.V. Sedelnikova, *The Lichens of the Western and Eastern Sayan* (Science, Novosibirsk, 2001)
15. *Red Book of the Krasnoyarsk Territory* (Krasnoyarsk, 2009)

# AUTHOR INDEX